THE EARTH'S CORE,
AN ENERGETIC COSMIC OBJECT
..
- A Geo-theoretical View of Matter Transformation -
Scientific Pamphlet

The essence of the theory of the Earth's Growth and Development.

Proposed by

Prof. Vedat Shehu, Ph.D., geologist and engineer

"The Earth's Essence is its energetic core, while core's essence is its transformable kernel"
Vedat Shehu

THE EARTH'S CORE, AN ENERGETIC COSMIC OBJECT

A Geo-theory of Matter Transformation
A Scientific Pamphlet

Vedat Shehu. Ph.D.
Professor Emeritus of Geology and Engineering
Faculty of Geology and Mining, Polytechnic University of Tirana, Albania

Scientific editor: Professor Clifford Ollier, University of Western Australia
English language editor: L. Joan Brown, Sharon MA Public Library, Literacy Volunteers of MA
Graphic Art: Leonard Gurabardhi

Cover:
First figure: Global tectonic rifts, borders of tectonic slices (plates); (from http://moodleshare.org/mod/page/view.php?id=5601)
Second Figure: Profile across Atlantic bottom (artistic presentation of the fig.2 of Pamphlet)

The Earth's Core, an Active Cosmic Object
Copyright © 2015

ISBN- 1512290874

Printed by CreateSpace, An Amazon.com Company
USA 2015

CONTENTS

Preface

In 2005 I wrote a preface to Vedat Shehu's previous book, *The Growing and Developing Earth*, and I now have the privilege of writing another one, as well as acting as scientific editor. Much of what I said then is still appropriate.

Shehu is a remarkable man with a rich background of experience. He was an engineering geologist working in the most practical and testable branch of geology firmly based on fact. You don't want your bridge to collapse! But he is also an imaginative dreamer who can think 'outside the square' and come up with novel ideas instead of following orthodox opinion. It is good to have your feet on the ground, as an engineering geologist, but there are times when you must abandon routine thought. Let's remember that Einstein said: "Imagination is more important than knowledge. For knowledge is limited to all we now know and understand, while imagination embraces the entire world, and all there ever will be to know and understand."

There are, of course, many problems when new ideas confront the orthodox, but orthodoxy cannot prevail. Fifty years ago we had different ideas in geology and cosmology from those of today, and fifty years from now there will certainly be new ones. So anybody holding all the conventional ideas of today will be wrong in the future, in some aspects. The problem for the truth-seeker is to know which ideas will prevail. Will Shehu's concepts prove true?

The previous book was clearly connected to the real world of geology, and I was able to point out the advantage that Shehu had in his Albanian background, which gave him a different viewpoint from many others who write about global geology. In the present book geography and local geology are of trivial interest. He has now stretched his ideas to cosmology and stars, and the whole Earth is just one example in the presentation of novel ideas.

Most of us live and think in a Newtonian universe with solid matter being affected in various ways by energy. Without giving the story away, I can say that Shehu is involved with such topics as plasma, the electric universe, and the transformation of energy into matter and vice versa, and he thinks the essential process takes place at the base of the Earth's outer core. On the way he has to explain grand ideas such as the formation of double stars, the origin of the Earth, and his views on the philosophy of the expanding universe. And all this in just 27 pages of NCGT Journal!

Of course others have ventured into this area and there are debates about whether we have a constant mass Earth and hence decreasing overall density as the Earth expands, or a near constant density as the Earth expands, or a near constant density and increasing mass Earth scenario. Carey[7a,b] proposed an increasing mass scenario and that energy entering the Earth from external sources condensed to form new matter leading to Earth Expansion. Maxlow and Eichler propose similar mechanisms, in which it is emphasized that a stream of charged particles (plasma) is being continuously expelled from the Sun. Some enter the Earth and form new matter causing the Earth to expand. According to Peratt, author of the text *The Physics of the Plasma Universe*, "Plasma is overwhelmingly the dominant constituent of the Universe. Yet most people are ignorant of plasmas."

Vedat Shehu has presented his own views. The reader of his booklet will be confronted by a host of novel ideas and will have much to think about. Does the concept of the Earth's Core Kernel provide a new outlook on the cosmos? I hope you enjoy reading this challenging book.

Cliff Ollier

Professor Cliff Ollier (Emeritus) is a geologist, geomorphologist, soil scientist, and honorary research fellow, at the School of Earth and Environmental Sciences at the The University of Western Australia, Nedlands, WA 6009

THE EARTH'S CORE, AN ENERGETIC COSMIC OBJECT

1. Introduction

"What would one tell about the peach's pit (core), if one knew only its peel and a bit from its pulp?.... Similarly, one might say that our planet did not make the core, but rather the core makes the Earth, our energetic Earth.' (Author)

Regardless of the fact that the idea of expansion does not fit with my concept of the growing Earth, as process of cosmic matter transformation inside the core, my article [38e] is ranked among the expansion theories, and about this topic was included in the section of Physics & Cosmology by the editor in chief[34] in the publication of selected articles from a scientific conference held in Sicily in October, 2011. The key concept of the theory of the Growing and Developing Earth is linked with two still unsolved problems: the first is what triggered the gaseous dusty cloud to become dynamic nebula of the planetary disc of solar system, while the second is what generates earth perpetual energy causing geodynamic processes and relevant phenomena, since the formation of the planetary system about 4.6 billion years ago. The Growing Earth theory from matter transformation solves both these problems and connects Earth's current development with its cosmic origin. The Earth's growth and development is such real phenomenon, as its inner energy generation is. Surely, if we should know how the Earth's inner energy is generated, then we should find how the Earth's growing process occurs.

It's known, the core is the essential part of every developing object having a core, including Earth. The aim of this scientific pamphlet is to argue that energy of geodynamic phenomena derives from the Earth's core, never less the theoreticians consider the earth energetic source the enclosed untouchable problem. So continually, are briefly present factual arguments, and laconically is demonstrated the truth that the Earth's energy cannot be generated in a scattered way through its heterogeneously differentiated geocentric structure of the silicate rigid mass; neither can it be generated from both parts of the core, nor from its insufficient radioactive isotopes. Naturally, this energy cannot be generated in any form from the center of the solid inner core. Hence, Earth's energy source has been determined as ultrathin geo-sphere, and named the *core kernel*. By its transformation, the particles and sub-particles are released from their ultradense state, and they form on one side fluxes of energetic radiation, and on another side they are reconstructed in nuclides of the common atom-molecular matter causing the growth of entire Earth' s structural geo-spheres: inner core, outer core, mantle, lithosphere, hydrosphere and atmosphere. This new way we are obliged to accept, and would connect the source of the Earth's geodynamic processes directly with the core's activity and its origin. Naturally, it seems unreasonable, because it doesn't conform to the current standard theory, by which the core is labeled as a plain heavier precipitate from the original agglomerated Earth.

This theory of mine starts from the universal reality that every kind of existence results from the previous species of existence through matter interaction and transformation and is mainly guided by two of my earlier basic geological principles:

(1) The Earth is a sample of the universe in my hands,
(2) The mater-energy unity of the universe is a
process of the energetic cosmic matter transformation
inside the earth's core.

Naturally, the energy release is the pivotal characteristic of the evolution of the planets and planetoids, and the intensity of their recent radiation depends on the current remaining unconsumed size of the core kernel of every single planet.

Furthermore, the scientific reasoning must respect some criteria. *Firstly,* it must distinguish between the consequence and the cause and, at the same time, it must avoid the manifestation of childish reasoning such as "When the tree leaves swing, the wind blows; when the leaves swing faster and faster, the wind becomes stronger and stronger; and when the leaves do stop swinging, the wind stops. It means that leaves cause the wind." Such reasoning is found in some scientific hypotheses when the premises are formulated with a lack of data, as for example: (1) "The earth's agglomerate mass makes the earth core," when it is dominantly an inert silicate mass, and the core has an energetic structure; (2) "The Sun originated as a hydrogen ball, and is releasing plasma," while the plasma of the Sun's atmosphere is a consequence of the transformation of matter somewhere in the Sun's interior. *Secondly,* it must determine that the unknown is inside the known and avoid baseless premises and the illusion of enigma and agnosticism. So, the known fact points to the unknown cause: the Earth and planets release energy, it is fact that points to unknown source of the earth's radiation, and Earth's growth. It is linked to another principle "from the particular to the general", and reversely it means from the Earth to Solar system and vice versa. The probable cause of a phenomenon must be determined by scientific abstraction. In our case, it is need to examine the inner planets' radiation and their magmatic activity as normal cosmic process of certain matter transformation, different from the transformation of the atom's radioactive nuclei.

Definitely, the theories of global tectonics, including the avoided Expansion of important authors[7],[31],[34],[35],[39] regardless of their trends and structural composition are interfacing with the cause of Earth's dynamism, a problem related to the cosmological factors of how the calm gaseous-dust cloud was turned into the dynamic nebula of the solar system of objects generating energy. Consequently, the missing mechanism of the Expanding Earth is related to unsolved problems of the Earth's energetic source.

I'd like to emphasize that this is evidence for, not proof of, a new form of matter in the core kernel, further, we shouldn't hesitate to reevaluate Wegener's phrase: "...all earth sciences must contribute evidence toward unveiling the state of our planet." In this brief passage, I am trying to bring in evidence for the essence of my theory presented in my book, "The Growing and Developing Earth" and in other articles[38a-e]. Let's continue the responding arguments along the following lines.

2. The Earth's Core within its Silicate Shell

The Earth's **core** (Fig.1) has its own entirely different geospheric structure, composition, and physical phase state, compared with its covering silicate shell (Fig. 6). The core silicate covering also has its own geospheric structure: the mantle and the lithosphere including its outermost geosphere, the crust. In the following lines of this scientific pamphlet is seen the author's consequent effort to shed light on the cause of such a radical difference between the core and its silicate covering referring mainly to the source of Earth's energy and related geodynamic phenomena. The phenomena and processes occurring in the Earth's interior are manifested on the crust's surface. So, the situation that continents are separated by the oceans and elevated above them does not represent the occasional surface position of the crust blocks above and beneath the sea level. It directly expresses Earth's inner processes causing two types of crusts:

> ➤ *Oceanic crust* of basalts and basaltic rocks composed dominantly by dark (melanocratic) Fe-Mg silicates (mainly olivine, pyroxenes, and amphiboles, serpentine) and partly by

feldspars (manly of the plagioclase) having higher density (3.0) and a lack of free silicon.

➤ *Continental crust of granites and granitic rocks* composed by white (leucocratic) Na- K- Ca silicates and aluminosilicates (mainly feldspars), and high content of free silicon (SiO_2) having lower density (2.7).

The Earth's inner processes are the cause of this difference between the crusts, and are related only to the inner Earth's structure, out of outer fluid geospheres. It's known, that all the events of Earth's history are recorded in the sedimentary stratigraphic sequences depicting Earth's evolution as if it were a gigantic book of stony sheets. The geologists and especially the paleontologists can read the *torn pages* of this remarkable *"stony book"* of the Earth's evolution.

The essential problem, waiting for resolution, remains in two questions: Is the core, with its amassing structure and behavior, a plain percolated precipitate, and can it stay in an inactive position inside the inert silicate shell originated from the accumulation of scattered gaseous-dust matter of a cosmic cloud? 2) Can the core, with its amazing structure and behavior, contain the energetic source? Let's continue to find answer.

3. Rocks, the Main Atom Molecular Shape of Matter on Earth and in the Cosmos, and Earth's energetic source

The atom elements free or in molecular bonds, including the elements of the rock minerals as well, are generated through nucleosynthesis in the stars, including the Sun and supernova explosion. This process, so called hydrogen burning, is interpreted as the fusion of hydrogen nuclei to helium, and, continuing as a chain reaction, other chemical elements are formed one by one, and simultaneously is released enormous energy. It is gigantic transformation process producing atom-molecular matter and energy. Many stars reject atom-molecular matter; others explode in the end of their life. In that way, gaseous dust clouds are formed, and the star-formation cycle starts within them. The galactic clouds and interplanetary space around the stars are full of grains, particles and frequently with the debris of rocky components. The silicate cover of the planets cores is composed of this material, and further with fluid one. The oxygen (O), this key element of everything, with the second dominant element, the silicon (Si), forms the quartz (SiO_2), a characteristic mineral of granitic rocks of the continental crust, and the predominant radicals [(SiO_3), (SiO_4)] of the rock minerals of the oceanic crust. When they are bound in one molecular bond with iron and magnesium (Fe, Mg), melanocratic (dark) silicate minerals of pyroxenes and olivine are formed. In these radicals some silicon atoms are replaced by aluminum (Al), and iron and magnesium are replaced by sodium (Na), potassium (K) and calcium (Ca) in different relations, leading to the formation of leucocratic (white) silicate and aluminosilicate minerals of feldspars. In addition, the feldspars and quartz (SiO_2) are main components of the the continental crust. These elements: O, Si, Mg, Fe, Na, K, Ca, Al, are the main and predominant rock-forming elements, building the whole rocky silicate cover of the Earth's core (more than 99%); they are inclined to form rocks and they are lithophile elements. Separated bonds of the metals with oxygen, sulfur, etc are differentiated separately and form rare concentrations of ore deposits. There in the rocks are even some other elements attached to them in negligible fractions. Even the fluid biophile elements (O, H, C, N) are, in different molecular bounds, common gases throughout the Cosmos. Also, many metallic elements enter into the biophile group. Some are needed for cells' vital functions (mainly iron Fe, copper Cu, Na, K, Mg, etc.) and others are needed for the construction of the skeletons and shells (mainly calcium Ca, phosphorus P, etc.).

Excluding the feldspar minerals present partly only in basic rocks (basaltic and gabbroic), the complex pyroxenes (Mg,Fe,Ca)2SiO4 and simpler olivine (Fe, Mg)2SiO4 are dominant components of the oceanic crust and the mantle (about 68% of the Earth's mass). The components of these minerals, even through widely different combinations and metamorphism (in serpentine and amphiboles), are dominant components of ultrabasic (mainly of peridotites, etc.) and basic rocks of mobile oceanic ridges[13], the Mediterranean nascent ocean[35e], and ophiolite belts of continents[14a,b]. These minerals form the predominating stony matter of the Earth, Moon, planets[23], asteroids, meteorites[45], the debris of comets, cosmic dust[51], etc. In addition, spectral studies have pointed out that in the space between the stars there are present grains of the pyroxenes, and olivine too[22]. Hence is concluded that rocky matter is formed through a universal mechanism of matter transformation related to atom-molecular formation and physical field generation.

Naturally, the scattered rocky materials rejected together with other molecular bonds from stars, was accumulated and gravitationally concentrated in well-known gaseous-dust clouds and further in the silicate cover of the planets. So, the problem stays to determine through the differences in the processes of the rock formation in the stars and in the planets, whether it is correct to consider rock formation in the stars, the last remnants of nucleosynthesis, and to consider it in planets, the last remnants of the planet's magmatic activity. It seems that the cause is the cosmic transformation of matter differentiated through the planet-formation process.

Rock formation and gas release as geologic and cosmic phenomena - The rock molecular bonds and molecular gases are the cosmic matter which composed the

Earth, planets, planetoids, asteroids and their ruins, and also compose the cosmic debris and dust particles originated from the stars. The growing Earth phenomenon must be understood as continuing atom-molecular additions to the Earth's silicate geosphere, predominantly rock mass. This phenomenon, as I think, might happen in two ways; (1) - by an unknown process inside the core attracting a kind of unknown radiation as ether[17], ether-like pre-matter[1] from outside and transforming it into common matter and energy, or (2) - by a certain ultradense transformable matter shape positioned between the cores[38a-e]. This second way, to the core-kernel's way of thinking, is correct, because such transformation generates physical fields and produces atom-molecular matter (a new way of the nucleosynthesis).

Surely, the ophiolite (ultrabasics) massifs are not scale aggregates, as they are interpreted by plate tectonics in the Alp-Himalaya movable belt[47], but rather they are protrusive penetration of viscous rigid mantle accompanied by basaltic lavas, consequences of growth process. There, in these cross sections (Fig. 3), the gradual formation of the ophiolite belt at the older crust fault, and as well in newer one is seen and indicates evidence of the growing process. Hence is argued that subduction, the 'vital' mechanism of Plate Tectonics, is negated by this fact. That was the essence of my advanced geological research[38] integrating many studies about the ophiolites of Albanides[14a,b], a small characteristic segment of the Alpine-Himalayan movable belt[40] which includes the Mediterranean as a nascent ocean[36]. The mobile ophiolite belts have accompanied the whole history of crust development showing an amazing similarity to each other and to them of the ancient Precambrian belts[5],[15].

Similar ophiolite structures also occupy the Caledonian belts such as the Appalachians[50] and the Urals[32], and even more special phenomenon is the similarity of the ophiolite belts to the recent generations of oceanic crust along the mid-oceanic ridge[13]. In addition, the Circum-Pacific and some other active volcanic zones are analogs of the ophiolite belts[35d,e],[25]. The expanding/growing Earth hypothesis also explicitly explains the presence of the continental rocks inside the ocean crust. Such phenomenon, identified in 85 cases[46] is used by proponents of continental crust *oceanization*[30] in fixed earth size and fixed continental crust.

Rock formation and gas generation as burning cosmic process - The stony matter, as seen in the rocks of planetary objects, in cosmic dust particles, in star-ejected material, and in supernova remnants, etc., is accompanied everywhere by gases and radiation[3]. An important indication of the inner origins of fluids is their presence in volcanic emanations. Knowing that volcanism, as a magmatic process, has been active throughout Earth's history, it is very natural to say that volcanoes have directly supplied the crust with rocks, the hydrosphere and atmosphere with fluids, and indirectly the biosphere with biophile elements.

The ash and fluids are last products of natural fire, after wood transformation through burning with oxygen interaction, in certain earth conditions. Similarly, the rocks and fluids are last products of cosmic burning, i.e., transformation of a certain ultradense matter shape in certain cosmic conditions after supernova explosion.

Let's see below meaningful similarities and differences of both processes of burning, according to the products after interaction and transformation.

$$\textit{Wood matter} \begin{cases} \text{ashes} \\ \text{gases} \\ \text{radiation} \end{cases}$$

$$\text{Ultradense matter} \begin{cases} atom - molecular\ matter\text{: rock minerals,} \\ \quad cosmic\ dust, grains, etc. \\ gases \\ radiation \end{cases}$$

It seems that the rocks and gases are the final product of the cosmic fuel ignition. The burning processes of both wood and cosmic fuel are the result of the different matter-shape transformation having certain similarities. From this schema, we must say that the remains of the burning process of both fuels: on one side dispersed remains of the wood (ashes, smoke, gases, radiation) could not produce again, directly by concentration, the wood, and as well on other side the dispersed remains of the cosmic burning processes, the gaseous-dust matter of clouds and related radiation, do not by themselves have the ability to concentrate and to form stars and planets with their cores. A certain mechanism must exist to recycle the burned matter. Ash and stone do not by themselves have the ability to concentrate and form the original matter.

According to standard explanations, all components of gas deposits in the Earth's depths, including He-3, are considered the product of degasification of the primordial rocky matter of the mantle that are released as volcanic gases. But, it is obvious, an energetic core produces fluids by the same transformation way they are formed in cosmos. Volcanism should not be a secondary process originated in the mantle; it is the primary planetary process of rock formation and gas release. The Earth's basaltic volcanism, after consolidation of the granitic crust of continents, became the dominant volcanism, and the cause of the differentiation of two types of the Earth's crust. Furthermore, newer crusts of the Moon and Mars are formed from basaltic volcanic rocks[23],[45]. That is why the continental shift is a phenomenon related to planet's origin.

4. Two Alternatives on Basaltic Crust Formation and Continental Shift

Key problem of two crust types - the outer geodynamic processes, naturally, do not influence direct the inner ones so, in this scientific pamphlet the sedimentary cover is not in consideration; the attention is focused only on the primary process of rock-formation, i.e. magmatic rocks of two global crusts, continental and oceanic ones.

The relations between both types of the Earth's crust are manifestation of the inner relations between the core and its cover. This became the pivotal point of discussions between different theoretical trends in global tectonics, which, as is well known, deals with geodynamics (i.e., Earth's dynamic structure in permanent change and development). Already, it has been clearly confirmed that the oceanic crust is newer, and it was formed step by step between the continents, while the continental crust slices were distanced from each other. This idea of mobile continents was named *mobilism, and since the beginning of the last century,* it has been expressed in two alternative notions, by constant Earth size[43] and by the growing Earth's radius[36]. Both alternatives of mobilism were rejected one by one, and the existing concept that continents are unmovable and fixed in their places was scientifically confirmed and *fixism* was dominant theory. But, in this Scientific Pamphlet the fixist concept of the continental crust oceanisation[30]in a fixed earth size is not taken into account. This discussion got new impetus, when the amazing data of ocean bottom confirmed those from continents. So, abandoned mobilism was revived as *neomobilism.* Let's briefly see common data of the divergent discussion of neomobilism.

Data from continents - The factual data that strongly demonstrate the fitting and shifting of the continental coasts are as follows: (1) The geomorphologic fitting; (2) The fitting of the rocks having the same composition and same age formation with the same fossil content, especially of land fossils; (3) The eroded and transported fragments from the base rocks of the one continent are found in newer sedimentary strata of another continent where respective rocks are absent in their base; (4) Some very rare magmatic rocks, *carbonatites,* are apparent in both coastal lines of the South Atlantic in the correct placement of their morphological fitting, and it almost seems as if they were cut by a saber into two separate pieces; (4) Sedimentary deposition of the same geological environment is present in many coastal lines of different continents.

Data from the oceanic bottom - These data are: (1) The bottom of all oceans is traversed by a global continuing mid-oceanic crack in the form of rift valley along mid oceanic ridge which, as a huge underwater mountain range is gradually rising some thousand meters from the plane floor, is facing the Himalayan ranges by size as a global mobile belt of the Earth's crust; (2) The central zones of the mid-oceanic ridges are the greatest mobile belts, where intense geological processes are running; the greatest heat flow, elevated frequency of earthquakes, volcanic activity and crust dislocation; (3) The entire ridge crest of basaltic rocks is of recent age, and there in the rift valley magmatic rocks are continually formed, and become new crust, while previously formed rocks are pushed further away producing a symmetrical positioning of the basaltic belts to either side of the global rifts; (4) Geologists have determined that rocks found in different parts of our planet with similar ages have the same polar magnetic orientation of the ferromagnetic micro-grains. Therefore, symmetrical basaltic belts of the same age, facing each other at equal distance from the rift, have the same polar magnetization, which was "fossilized" along the time that the Earth *reversed* its magnetic field. It is a fact that the age of the basalt belts of the ocean bottom is increasing from the ridge rift to both continental coasts, where the first basalts formed along both sides of the global original crack separating previous unique continental crust; (5) Volcanic and seismic activity also occur in some peripheral long, narrow and deep geomorphologic structures named *oceanic trenches*, which are situated along *island arcs* in the marginal arc of the Western Pacific.

Finally, these observations are indispensable facts that the two phenomena are occurring simultaneously: 1) The continents are going away from each other, and (2) the ocean crust is growing. So, the hypothesis of *the continental shift,* rejected in the past, *already has become a fact*: The continents are going away from each other but together with oceanic crust generated along faults of middle oceanic ridges (Fig. 2)[38e]. This situation puts forward a big planetary problem to be solved: How was this additional gigantic basaltic crust formed? Was it formed in constant earth size through the circulation of crust and entire lithosphere, or is it evidence that argues in favor of the growing earth process through the *core kernel* transformation? Actually, the theories of the Global Tectonics are classified in two gropes[28]: *firstly,* the *Fixed-Earth theories* which assume the Earth has maintained an essentially fixed radius, here, together with continental crust *basification,* i.e, *Oceanization,* is included even *Plate Tectonics,* and *secondly,* the *Expanding-Earth theories* which assume that the Earth's radius has increased over geological development. *But here in this grope is not mentioned specificity* of theory of this Scientific Pamphlet, *The Growing and Developing Earth* by core kernel transformation since the planetary disc inception.

We are now staying between the standardized interpretation of *Plate Tectonics* and its rejected alternative known as *Earth Expansion.* Factual base of both theories can unified by same source generating the energy for geodynamic phenomena and atom-molecular matter added to the growing process. We must discover how it occurs.

Plate Tectonics - Naturally, according to the previous level of cognition, any alternative of continental shift was unacceptable because of the absence of a process to cause it. Hence, at that time, the expanding earth alternative[36] was rejected as impossible by Wegener's interpretation[43]. Later, Wegener's variant was rejected by scientific authority, because it did not have an acceptable argument or mechanism to cause this phenomenon. The new discovery, the separation of Earth crust into global tectonic slices by global faults, assumed that the crust is divided into *tectonic plates*. This newly known reality was given the new name of abandoned mobilism, firstly known as *neomobilism* and then as *Plate Tectonics*.

We must be aware that the movement of the *tectonic plates* (slabs) is objective reality and not a function of the mechanism of Plate Tectonics. The founders and consequent followers of the Plate Tectonics theory[10] rejected the Expansion alternative categorically, and, consequently, they became witnesses of the standardized cosmological theory of fixed Earth size. Therefore, the theoreticians of that time enthusiastically announced that Alfred Wegener's theory[43] is true: the ancient unique continental crust was broken into separate continents, and with respective growing oceanic crust, have distanced apart to their current positions in constant Earth size. The founders of this theory were obliged to create and strongly support three mechanisms:

- *A radioactive decay of elements* such as Uranium, Thorium, and an isotope of Potassium are the source of Earth's energy.

- The *subduction mechanism*, by which it is imagined that the flowing circulation of the upper mantle (the so-called fluid asthenosphere) , causes circulation of the entire thick lithosphere (crust) in the same rock mass as it is added to the mid oceanic ridges.

- *Convection flows of the mantle* capable to drive and circulate the thick lithosphere.

In fact, these three processes currently occur on a minor scale, and they are maximally exaggerated in accordance with the standardized cosmological theory of the constant size of the Earth. In the early phase, recycling of the lithosphere was a normal phenomenon, but gradually, simultaneously with the growing process and as the crust distanced itself from the source of energy generation (the core), the effect of convection was decelerated until it stopped to play this role[38a-e].

The Expanding Earth - This theory is based mainly on the same data and evidences as Plate Tectonics. In addition to this, the disdained expansion phenomenon was proved by an interesting test[17] that demonstrates that the continents fit together perfectly and envelop the Earth with continental crust on a globe some 0.5 to 0.6 of its present radius[39],[42]. The expansion is one of the only two possible ways to explain the Earth's mobilism. Expansion is generally refused because of the lack of a mechanism to bring it about; just as Wegener's Continental Drift hypothesis was rejected for lack of a mechanism. Similarly, the lack of arguments for a real mechanism is not only a defect of the Expansion theory, but of all *trends* of the global tectonic theories. Many data have been brought forth and many interpretations have been written about Earth expansion (growth) using the factual data and evidence of Plate Tectonics, but rejecting subduction as an unfitting ornament above the hidden truth.

The Earth's core and expansion - As it is seen, Earth Expansion remains the unique independent alternative to explain the Earth's enlargement by the gradual formation of the igneous basaltic crust among the continents which continue to move apart in same crust bloc with respective ocean crust. Hence, the theoretical duty is not to bring another facts to face up to the Plate Tectonics and to support Expansion, but to reassess and unify the current hypotheses of the Earth's-energetic and igneous basaltic source, and to reevaluate it as a consequence of certain process of matter transformation in the interior of the Earth's core.

Many authors are already connecting Earth's energy and expansion with the core or with the Earth's center, but aside of the concept of the matter transformation producing energy and atom nuclei. Among earlier authors, an interesting view is presented by Hilgenberg[17] on the role of unidentified space ether that intrudes into the centers of cosmic bodies where it is transformed into ordinary matter. In this way, the Earth is enlarging uninterruptedly in the same way other cosmic bodies are enlarging. Similarly, the idea of unceasing expansion was developed as an important theory during the entire last half of the twentieth century in different books, articles and activities of Professor Samuel Carey[7], and it is currently advanced by Professor Giancarlo Scalera[34,35a,b] and his colleagues, [Ollier[29 a,b], Maxlow[24 a,b], and Cwojdzinski[11]]. At the same time, the possible cause is left aside of the geosciences as a problem belonging to cosmic sciences. Other trends of a continuing expansion are associated with cosmic mass accretion and cosmic radiation [Blinov[6a,b] and Myers[26 a,b]]. Similarly, a specific hypothesis of growing Earth is currently proposed by Adams[1], reflecting that the spontaneous *matter creation is caused by some sort of electron-positron pair production,* which creates every cosmic crystal grain. Further, crystal is grown in a meteor, asteroid, planetoid, planet, giant planet, star, and in every galaxy throughout the universe, causing particles to grow *permanently and endlessly.* Then let's consider the confirmation of Earth's enlargement in sequential studies[28] with different concepts under the notion of *expansion*[19], or *growth of a superdense inner core.* Commonly, the expansion proponents view the whole core[12] or only the inner core[31] as the cause of the process. All the different interpretations of the core's role in the expansion process are already integrated by Scalera[35a] in two main variants: (1) the Earth was originally an ultradense unique core, and it started to form Earth's current structure through spontaneous transformation into common matter, and process will continue until it becomes exhausted causing the Earth's expansion to stop; (2) initial small Earth began with its current structure, but with an ultradense inner core, which is expanding together with the entire Earth structure, and it will be permanently expanding into a giant planet reaching the size and activity of a star. It is also confirmed by even more current information from scientific observations that the mantle rather core's silicate shall does not contain the source of the Earth's energy, the generator of the geodynamic phenomena.

Some papers are in quandary to seek for a new energetic source[41], and lately assess and reassess the impossibility of both mobile theories, and while correctly reject the coined mechanism of the favorite hypothesis of Plate Tectonics[41a], are trying to decorate the older coined mechanism, oceanisation[30].

The Growing and Developing Earth - It is my theory that a core-kernel (so named by me) has been the cause of the generation of both the energy and of the atom-molecular matter as additional mass for the basaltic crust and the whole Earth growth. Therefore, the Expanding Earth, with its core kernel mechanism, would become The Growing Earth. The aim of this Scientific Pamphlet is not to add more to this question, but to solve the problem of the growing process, to offer an explanation *not about why, but about how*, the so-called expansion occurs as a phenomenon of a certain process, not as unknown property drifting in from the Cosmos.

The *why* of the problem is solved simply by one fact that the reality was wrongly interpreted by Plate Tectonics (fig. 6)[38e]. The Plate Tectonics theory has been modified many times and eventually became plume plate tectonics. It cannot bear more modifications.

If Plate Tectonics were to undergo another modification it would have to accept that ocean bottom is not spread but grown as evidence of the whole Earth growth, and it would become a theory of *Tectonic Plates of the Growing Earth.*

The Earth's energy source, the core kernel mechanism makes the core an energetic cosmic object; this differs basically *the Growing Earth* from *the Expanding Earth.*

Earth Growth, not Earth Expansion. Why? The essence of "The Expanding Earth" theory is the enlargement of the Earth's radius, and if we determined the source of this phenomenon i.e., energy release and silicate rock formation, then Earth expansion would become Earth Expansion through a Growing Process and would converge with "The Growing and Developing Earth" a concept of mine (fig. 4). However, there is a *distinct difference between these notions, expansion and growth.* Let's examine this difference[38e]. When a fruit is getting ripe it is enlarging its radius, while the fruit itself does not expand but it grows. What happens if we cut an orange rind, still green, up to the pulp with a razor blade leaving almost an invisible line trace? When the fruit ripens, we will see that a furrow is open along the line (Fig. 4A). The same phenomenon occurs also if we strike a young tree trunk with a hatchet cutting its bark. Again, an almost invisible trace is formed. It seems that such a "wound" would close, but after many years the line will become a furrow (Fig. 4B). In both cases the environmental matter of the ground and air oxygen, under photon radiation has been transformed into cellulose and other relevant organic substances, which feed and grow tree tissues and tree fruits, but the newly transformed matter did not circulate between the partitions of the cuts, and the partitions did not connect with one another, but simultaneously the partitions moved apart along the tree growth. Similarly, by the growing process, the furrows are formed spontaneously in the dead bark of older tree trunks (Fig. 4D).

The notion of the expansion might be illustrated by the moving apart of the pieces on a ball after it has been inflated (Fig. 4C). Conclusively, the expansion does not contain the notion of the growing process resulting from matter interaction and transformation. Apparently to us, the expansion and growth appear to be synonymous, but as it seems, Earth expansion could not be an inflation (Fig. 4C), but an enlargement through a process of growth (Fig. 3A, B, D). Earth is growing by its own inner matter transformation, while outer radiation might play a similar role as electromagnetic (photon) radiation in growing plants, not influencing directly the mass addition from the growth process.

If, the Earth energy would be evaluated by us, not something in itself but, as it is in fact, the energetic fluxes, which are generated together with sub-elementary particles of the atom's nuclides; then we should conclude that cause of divergent interpretations in the global tectonic theories should be eliminated, and different interpretations should be unified under concept *"Tectonic Plates of the Growing Earth"*. So the term "expansion" was replaced by real growth process, and the reality that Earth's crust is broken in the tectonic slabs or plates, is used, and left aside the imaginary mechanism, subduction of Plate Tectonics.

5. The Earth's Core, an energetic Cosmic Object

"The core of the Earth is a miniature sun inside the silicate cover, while the Sun is a gigantic bare core." (Author)

Is the core not only the central part, but, as a sample of cosmic matter transformation, the most essential part of the Earth? The answer to this question belongs to the whole Scientific Pamphlet, and in this paragraph I present the essence of the theory.

Energetic essence of the core of the Earth - A nut, an almond, is a natural core (pit) with its kernel inside. The core (pit) of a peach fruit is similar to the almond nut, while the pulp is covered by the peel. The kernel of the core is its

25

essence, its seed, containing the embryo of the growing process of the flowering trees. The embryo, in all kinds of things, is the result of matter interaction and transformation. Naturally, the fruit core-kernel is a specific kind of organic matter, which results from matter interaction and transformation, and simultaneously it does inherit the property and trend of matter interaction and transformation. So, the kernel of the peach's core is built by certain transformable kinds of vital bioactive molecular bonds. It makes the core the essential part of the fruit, because through interaction and transformations it does generate the energetic biogenic molecular bounds for the embryo sprouting and seedling growth leading to the cyclic processes: formation of the tree, and further peach fruit. Similarly, with its property of matter transformation, the core of the Earth contains its transformable kernel. The kernel of the Earth's core is built by a certain, pre-atom-molecular shape of matter, Then, the core-kernel is the essential part of the core, and makes the core the essential part of the Earth. Hence I formulated the essential meaning of my theory plainly in this way: *"What would one tell about the peach pit (core), if one knew only about its peel and a bit from its pulp? ---One might say that our planet did not make the core, but the core, rather the core's kernel makes the Earth, our dynamic Earth."* Then, to me, the mantle around the core resembles peach pulp, and the lithosphere resembles the peach peel or orange bark.

Conclusively, *the core of the Earth is a miniature sun inside the silicate cover, while the Sun is a gigantic bare core.* If it is so, we can transit this problem step by step from the particular to the general, from the Earth's geo-dynamic processes to the Sun's energy; from global tectonics to cosmic matter transformation and the origin of the Earth's core, furthermore, from the Earth's core energy to beyond the Bing Bang's cosmos of the expanding sky. However, we stop only to consider why the source of Earth's dynamism is inside the core. Then, proceeding so, we must investigate the core's specific concentric-spherical structure, which has to demonstrate that an energetic Earth source works inside it.

Here in the following explanation, are reviewed the observed facts and respective interpretations on Earth, which are in conformity with fundamental hypotheses of the Earth and cosmic ones, as well. Almost all these interpretations adapt the facts to the standard theory of Earth core inactivity. Any proposed mechanism[37] does not justify the separation and percolation "downward" the iron and nickel from the initially accreted agglomerate material, if it neglect to explain the core energetic structure.

The Simplicity of the theory of the core's energetic source - This theory is propounded by me as *a geo-theoretical view of matter transformation, or* shortly *Geotheory.* It is widely known that everything is a result of matter transformation, including Earth's inner energy. The standardized theory of the source of Earth's dynamism, its energetic source is still a matter of discussion about what triggered the original nebula to become our solar system. However, in Earth's interiors, the intentional observations concluded, the common chemical atom-molecular reactions cannot generate the required energy. In addition, the release of energy through the disintegration of certain radioactive isotopes is insufficient to be the only energy source. Nor could the imagined original gravitational collapse have been the cause of a permanent source of energy. Thus, naturally is needed to conclude that *inside the core is positioned a certain cosmic matter which is spontaneously disintegrated, and causes Earth's energy and Earth's growth.* Hence, it should be understood that the core was an innate, energetic cosmic object inside the Earth since the epoch of nebula transformation in planetary disc.

How might this happen? It is plainly a natural process that could easily be captured by proper and simple scientific integration of a few common, well-known facts which are essential to the process. This way of simple scientific thinking was applied by an ancient scientist, Anaximander

(610 - 546 BCE) in his forgotten definition about natural biological evolution. Even more so, the forgotten heliocentric model proposed by another ancient scientist, Aristarchus (ca. 310 BC – 230 BC), also speaks to this simplicity issue. Four centuries later, this model was totally ignored by complicated mathematical accounts of Ptolemy (90 – 180 CE). So, I am obliged to emphasize that, in writing this compact document, it is not possible to refer to all the details of well-known publications by numerous authors about discussable problems of standard theories. In any case, these details would not bring something new to shed more light on this question.

The Earth's core, a sun-like star in miniature - Let's see and analyze the detailed design of the Earth's core shown in the attached Figure 1. This is a scan of the core's tomography which geophysicists have taken from earthquake waves. Here I have found the undetected mechanism for this theory of the core as an active cosmic object. So, it becomes necessary to shift the source of the geodynamic processes from the innert mantle to the dynamic structure of the core. Following explanation, we will then say: it is as plain as *Columbus's egg*. Here, I am underlying again, I have found the undetected mechanism, the pivot of this theory. An important phenomenon discovered by the geophysicists is that the full rotation (in 24 hours) of the inner core is done differently from a full rotation of the outer core and the whole Earth. This is one of the phenomena, impossible to be captured by the scanning, but it points to core energetic structure, and to its function. In Figure 1, core energetic structure is demonstratively expressed by the dominant position of the melted outer core within two rigid geo-spheres; inside the silicate mantle, and outside the solid inner core. Furthermore, it is reinforced by the fact that the boundary of the outer core to each solid geo-sphere is built by a narrow transition geo-spherical zone (layer) composed with mixed material; melted and solid. These layers point to gradual transition of melted matter to solid matter. Then, if this solidifying mass of outer core had not been constantly generated, the entire core would have stiffened long ago. Therefore, such function of core structure does directly repudiate plain Fe-Ni composition and inactivity of the core. In this way, the core energy release is plainly confirmed as occurring through a certain matter transformation; surely, not through radioactive disintegration in the core, but a natural cosmic process is taking place there, between both solid geo-spheres.

In the process of trying to unify Earth's geodynamic processes with the growing (spreading) oceanic crust, I have reassessed, from the viewpoint of the core energetic source, standard interpretations and old and modern discoveries inside the Earth and in the Cosmos. I have, therefore, formulated, as a scientific truth, the existence of an ultradense energetic source, a focus, positioned as ultrathin geo-sphere similar to a diaphragm between both cores, drawn as a full circular line in figure 1 and as an interrupted line in figure 6. This focus, having the function of the earth growing process, is named *the core kernel*. The core kernel mechanism makes the core an energetic cosmic object having a rocky silicate cover. Of course, the ultradense core kernel, by spontaneous disintegration, causes, firstly, Earth's energy and, secondly, formation of chemical elements in Earth's growing process. This process of formation of the nuclei of chemical elements and generation of energy occurs in the Sun, as a star, and in supernova explosion, as well. Then it means a miniature similar process of nucleosynthesis occurs in the core, but surely the starting point of this proces is different from the way standard theory explains it.

Core kernel supporting arguments - If we can understand the necessity to shift the energy source from the mantle to the core, then it is necessary to create a new concept about the core's role in Earth dynamism since its origin. Now, let's proceed to determine the cause of the energetic source within the core; some facts as effects are pointing out to the cause. These facts and evidences are as follows:

1. The melted state of the outer core, in contrast with the solid inner one, points directly to the cause indicating the core's dynamic structure.

2. Two transition zones (of the mixed phase, melted and solid layers - D" and F (in Figure 1), from the outer core to both rigid geospheres (mantle D and inner core G), also point to a probable energetic focus between the cores, which generates atom-molecular matter going toward both sides and which are gradually solidified.

3. Well-known differences between the outer core and inner core, especially the differentiated rotation of the inner core from that of the outer core an energetic boundary such as a geospheric screen that causes such difference.

4. Intensity of the Earth's phenomena such as seismicity, volcanism, and geomagnetism concur with the intensity of the Sun's activity. This speaks to the fact that the processes occurring in the Sun resonate with those occurring in the Earth's core, as processes of the same origin.

5. The so-called spreading of the ocean floor is really ocean crust growth with newly formed stony silicate matter as evidence of entire Earth growth. (See Fig. 2 about oceanic crust growth) This formulation, as interpreted by me, points to the fall of the hypothesis of subduction mechanism and confirms the necessity of postulating a growing Earth's energetic mechanism, the core- kernel.

6. Hotter plume structures, moving from the core-mantle boundary outward, illustrate the paths by which the newly generated atom-molecular matter, accompanied by radiating energy, moves through the mantle within an inert, relatively colder environment. This phenomenon naturally negates the subduction of plate tectonics and confirms the necessity of postulating a growing Earth energetic mechanism inside the core, the core kernel.

7. The content of helium in lava (predominantly He-4,

and a fraction of He-3) and in the deep and great gas deposits (sometimes up to 7%) is significant evidence of the inner generation of chemical elements and their bonds, including natural gases. Furthermore, while He-4 is released from both the Sun and from radioactive decay, helium-3 is produced only in the Sun. Then its presence in the depth of the Earth shows that it is probably synthesized in the Earth's core, as in the Sun, .

8. The explosive supernova nucleosyntheses probably continues through its ultradense remnants present in the solar-forming nebula. These remnants may cause the gravitational collapse of the nebula (Fig. 7).

9. The current discovery of the rogue planets orbiting galaxy, and planets orbiting a binary star (twin sun-like stars), is demonstrative evidence of the core kernel mechanism, where in on side causes formation of the remote planets, and in another side causes formation of the smaller sun like satellite of the larger sun. It comes directly from a specific accretion of the ultradense kernels, which leads to the formation of a minor satellite star instead of a core of a gigantic planet within the cosmic nebula (Fig. 7).

Core kernel, an ultra dense cosmic matter shape - The core kernel theory cannot include concept of the *dark matter* which absorbs electromagnetic specter; it still stays not identified by the structure, and; in particle physics is not yet characterized type of its subatomic particles[9], their probable function, and if the supposed dark matter can influence star-planet relations. Naturally, core kernel notion is a derivate of the cosmic ultradense matter and explains a new circulation way of visible matter of the electromagnetic specter.

The old definition against *matter-energy unity* is still wrongly in use. Such definition is limited only to the phases of the atom-molecular shape of matter: solid, liquid and gas, even adding plasma as overheated gas, a state of heterogeneous atomic matter discovered later. The core kernel transformation is in accordance with the confirmation in physics of the Einstein's determination that "energy is the same as mass" (expressed by his famous formula $E = mc^2$). Just, my theory that the very source of Earth's energy takes place by means of a certain ultradense matter transformation in the core (more correctly, in an energetic source of the core), confirms this formula of the matter-energy unity. This unified concept of matter shapes is, as well, in accordance with further studies in quantum physics, the atom's nuclear structure[18], and in conformity with the theory of cosmic matter circulation. Obviously, from this real concept of matter is actually reached formulation that the basic matter shapes are formed by concentration and disintegration of the particles and ultra-particles. Furthermore actually is put forward whether the particle fluxes, such as quarks and gravitons (still unidentified), may be concentrated in ultradense cosmic bodies of a black hole type, and afterwards may happen the disintegration by a blast, blast of a supernova type, and the new cycle of the interaction and transformation has to begin. Furthermore, this concept of the transformation of the ultradense matter shape in particles and ultra-particles is in accordance with experimental observation by the most powerful particle collider of individual quantum particles being just particles of the physical field matter shape. So, all matter types and states known in physics are included in the matter notion of the following shapes, in permanent interaction and transformation:

- The ultradense shape of matter of some compact stars and of the atom nuclei is subject for astrophysics and nuclear physics;
- The ultra-rarified shape of matter, i.e. wave–corpuscular shape (physical fields, radiations) is subject for field physics and particle physics;
- The atomic-molecular shape, i.e. common matter is a subject for chemistry;
  Plasma is a transitory unstable state of the pre-atom-molecular shape of matter which transits into

radiation and chemical elements and their molecules; and it is a subject for plasma physics.

In this concept about matter's shapes we are able to accept their energetic release from the core kernel, the cause of the growing Earth. When arranged in atom-molecular shape, nearly the same mass of the particles and ultra-particles disintegrated from the ultradense core kernel, does occupy a considerably greater volume. As a consequence of this, the Earth does grow. On a global scale, the growing Earth process is schematically presented in Fig. 6, where it currently shows the achieved structure of Earth growth from the inside and the outside of the core kernel, and using well-known rifts, as illustrated in Figs. 1 and 2. What kind of particles are in super-dense state would be in the range of the images of the theoreticians of astrophysics; perhaps this state of the matter is of the nucleon's constructive ultra-elementary particle, the quarks. Probably the quarks have ability to interact, create atom nuclei and emit graviton fluxes of the gravitational field. Hence might be supposed the gravitization process of the objects having not transformable core-kernel. Space cannot be empty; if a vessel is empty of water is full of air, if a space is empty of atom-molecular matter is full of particles, ultra-particles and wave-corpuscular fluxes or of *hypothetic ether;* vacuum is nonexistent [36c].

Calculation of the core kernel - Already is argued that ultradense core kernel occupies a certain mass of the core calculated as to be compressed only of atom–molecular matter. A new criterion for evaluating the mantle's and core's density enables us to calculate the mass of the core kernel. Let's do reassess the density calculation through the known velocity of the seismic p-waves across earth structure. According to the standard calculation, the core of the heaviest fraction and under highest pressure is of very high density, about four times greater than crust density. The core kernel being ultrathin is indiscernible by the seismic wave velocity. Here, according to the functional dependence of the velocity of seismic waves (especially p-wave velocity) on density, there is an observed discrepancy in the principle of the calculation of matter density for the inner Earth's spheres. At the Earth's core, longitudinal wave velocities (the cross waves die away) are lower than those of the outer mantle and lower still compared to those of the inner mantle and specifically of the D" layer. In order to calculate the density of the inner Earth's shells and of the Earth's core, new criteria must be chosen. Due to the well-known fact that the Earth's mantle is composed of rocky matter that is in a rigid to a viscous-plastic state, its density should not be greater than 3.9 - 4.9 g/cm^3 contrary to earlier calculations of 4.7 – 5.7 g/cm^3 (even compared with Mars with its density near 4.0 g/cm^3).. On the other hand, regarding the outer core in its melted state, we must not take a much higher value than the maximal value of the mantle, perhaps around 4.9 g/cm^3 compared again to the commonly proposed value of 9.9 g/cm^3. Even the inner core density has to be around 6.1 g/cm^3 contrary to earlier calculations of 12.2 g/cm^3. Reduction to these new density values would reduce the core's mass by nearly one-third to one-half, and the Earth's mass by 10 to 15%. The mass of the core kernel consisting of super-dense matter shape might then constitute at least 10% of the total mass of the Earth.

6. The Kernel of the Earth's Core, the Generator of the Geodynamic Processes

Logically, inner energy generation is accepted as the cause of the inner geodynamic processes, which are expressed on the surface by geodynamic phenomena: earth tremors, crust fissures, lava expulsion, and crust tectonic displacement. These phenomena are studied by global tectonics (geotectonic theories). As it seems to me, global tectonics is becoming an essential part of planetary science, because it already aims to relate Earth's geodynamic phenomena and processes with other planetary objects of common origin in the solar system. Furthermore, it does study the dynamic development of our planet in a solar system in our galaxy, and, as it is well known, our Milky Way is one of the uncounted galaxies in space spread limitlessly even beyond the reached boundary of our radius of observation. It is becoming an interdisciplinary field, growing originally from geology to earth science, but which now, together with geochemistry and geophysics and other disciplines, is oriented to planetary geology, and it incorporates planetary cosmology.

Geodynamic phenomena - Earthquakes, volcanoes, and tectonic motion and deformation are different consequences of the same cause – the processes of the growing Earth. Among these geodynamic features is a functional space relation demonstrated by their common distribution originally along the global tectonic rifts[36b], particularly, inside the crust's mobile belts and along the edges of tectonic slabs (plates): through the middle oceanic ridges, the Pacific Ring of Fire, the Alpine orogenic system, and along some recent active tectonic faults, especially the East African Rift and Sun Andrea Fault. These phenomena manifest Earth's dynamic processes, and do not cause one another, but rather are all independent effects of a common cause, and only a sub-set of them can cause the volcano-seismic event[38a-e]. In Figure 5, is interpreted the seismic tomography of the mantle in relation to its manifestations in the outer stony geosphere (lithosphere including crust), and it tries to give a clear panorama of the mantle lateral heterogeneity and the radial plume structure. The plumes[33],[48] are identified seismically among relatively colder parts as hotter pillar-like regions of the mantle. The plumes are evidence of the motion of the hotter silicate matter from the core–mantle boundary outward. This trend of the motion does not show the interpretation of the Plate Tectonics supporters who suppose inward motion of the

35

cool lithosphere, reheated when it reaches the core, and returned as hot plume. On the contrary, it demonstrates that the new hotter matter of the plumes moves outward inside the less hot environment of the mantle and has no return back to the core. By this way, throughout the process of Earth development the plume structures are formed[33]. The plume flows start from the core-mantle transitory zone (D" geospheric layer), and reach shallow active hotspots of the volcanic centers, causing tectonic stresses and tectonic movement, and both demonstrative phenomena: earthquakes and volcanoes[8a,b].

Earthquake shock and tectonic deformation - Well-known seismologists are already reassessing the new discoveries attributing the inner energy to the core "as an energy source for tectonic and magnetic activities in the shallow Earth..... The fracture zones are channels for the core-derived energy to rise to the shallow Earth.... The energies are stored in the nearest available space." [8b,c] This certain mechanism as an energy source, and a source of matter transformation inside the core. Furthermore, if the source of energy is related to matter-energy unity, then the source of matter transformation releases energy together with atom-molecular matter as pre-magmatic matter. So, it would be the fact that matter and energy are generated from core kernel transformation, and both move together from the core outward along 'channels', or spread as plumes through the mantle. A certain part of this plume mass along its way upward stops at different levels and increases the viscous–plastic mantle. While the hot mass addition in the mantle is gradually interfacing with the mantle's viscous state, but it is concentrated, solidified, and it creates gravitational disequilibrium of the geoid in the solid, cold lithosphere. Then, when the equilibrium is replaced, the crust is shaken or shifted, causing earthquakes or tectonic deformation and displacement.

Therefore, when somebody speaks about seismic energy accumulation, this also means that some mass was generated, shifted and accumulated together with its potential energy, deforming the Earth's spheroid gravitational equilibrium. In this case, the gravitational field acts to replace the deformed gravitational balance in the sphere causing slow tectonic movements and those occurring suddenly as earthquakes. The earthquake foci are determined by the lithosphere-mantle boundary. Seismic tomography points out that the lithosphere's bottom is highly undulated; parts of the brittle elastic lithosphere are sunken deep into the viscous flowing mantle.

Hence, when matter, on its way upwards, is accumulated in bulk, the spherical balance of gravity is restored at a deeper level with stronger intensity. A smaller mass of matter on its way to the surface creates a gravitational sphere disturbance nearer to the Earth's surface, where the process of restoring balance occurs with softer intensity. Earthquakes are also connected to Sun-Moon position according to the site of the earthquake and are triggered by related Earth tides[20]. Perhaps, the gravity of the Moon and Sun accelerate this previously prepared inner Earth event.

The events preceding the earthquake are emanations of the He-3 isotope and the disturbance of the local magnetic field at the earthquake's epicenter. These phenomena might be used as diagnostic signs to predict an earthquake.

7. Core of the Earth and of the solar system bodies

In this scientific pamphlet, the Earth's energetic source is presented as already proven, and calculated that occupies more than one tenth (>10%) of the Earth mass; its composition is determined as concentration of particles and ultra-particles in ultradense state within the core. This energetic source is just the ultradense matter of the transformable *core kernel* (see section 4, above). .

The Earth's core is not an exception but only one case

between others, resulting from the same process, which led to the formation of the planets in different orbital conditions. Besides the orbital conditions, the notable difference among planets and other round objects is related to core kernel sizes and the attained levels of their transformations, while the Sun differs as a star of his group. The objects without a round shape did not result from any energetic source; are different rocky fragments remaining out of stunned planet-formation process, or are ruins of potential planets.

Earth core dynamic structure - As mentioned above, some new interpretations, modifications of standard ideas, give the core a limited role of energy generation. These interpretations are attracting attention by pointing to the function of a probable energetic mechanism within the mass of the outer core. Naturally, the core energy is not a virtue of one condition, but is the result of the planetary processes responding to circumstances.

As we carefully observe the structure of the Earth's core (Fig. 1), we can distinguish that it is an expression of an energetic structure due to core kernel transformation. The position of this energetic source determines the Earth's growth process to occur on both sides of the geospheric core kernel, outward and inward. Naturally, dominant part of the transformed atom-molecular matter and of the energetic fluxes flow outward. Probably, a smaller, denser fraction of matter and smaller doses of radiation are involved in the inward motion, supplying the inner core with new matter and energy, causing its solid state and its specific surface morphology. Surely, newly generated atom molecular matter has tendency to be differentiated and stratified according to density. It points to a thread that connects the core activity and the Earth's cosmic origin shared by other planets, which have to be explained by the same universal processes of matter transformation. All planets and planetoids have such a structure with differences caused by the size and the transformation level of the core kernel. It also confirms the core kernel's role in the planet formation process.

Also, the cores of the other planets couldn't form as the result of inert precipitants of denser components in the center of silicate agglomerate mass. The essence of their cores, similarly to that of Earth, was the transformable core kernel. The growth of every planetary body occurs by such transformation.

The cores in the dwarf planets - Our Moon, as well as the asteroids Ceres and Vesta, the moons of the giant planets and the planetoid, Pluto, are all well studied in detail and determined as dwarf planets. Furthermore, many similar dwarf planets are discovering beyond Pluto's orbit. These are stagnated potential planets after their energetic source exhausted, each generally having a uniform core and no sign of inner activity. It appears that very early in each case, the core kernel has been totally transformed. They became dead bodies each having a differentiated core, mantle, and crust with demonstrable evidence of a minor ancient volcanic activity.

Energetic core in an exceptional moon - In some moons, geological activity is still alive in different phases. In Jupiter's Io it is of maximal intensity. With over 400 active volcanoes, Io is the most geologically active object in the Solar System. That points toward the existence of an energetic core kernel, which opposes the current interpretation about the alleged presence of radioactive elements in large quantity.

Our Moon might have been in a phase like this of Jupiter's Io nearly 3.3 billion years ago. The volcanic activity and the formation of the rocky matter in the Moon has continued, after formation, for a period of more than 1.2 billion years (the age difference between asteroids and moon basalts) and, at depths, it still continues today evidenced by the presence of volcanic gases and seismic shocks. This shows that the core kernel in the Moon is in the final process of total exhaustion, but it is still not a dead body.

The cores of the inner planets - Inner Planets clearly show differences in the processes of their core kernel transformations. Venus and Mars have cores in proportions to their sizes, and while historically, also earthquakes are not recorded. Perhaps do exist supposedly as rare phenomena. This points to a very limited size of the kernel that remains untransformed. Core activity of these planets as been very active through their historic development it is demonstrated by dead large and extra large volcanoes, especially in Mars. Mercury has specific large core partly peripherally molten pointing to the current still active core kernel.

The cores in the outer planets - Outer planets are giant planets; they previously were considered to be entirely constructed by gases H2, He, CH4, NH3, H2O, but later has been discovered the strong inner radiations and strong magnetic field in every giant planet. Among the planets, Jupiter has strongest and largest magnetosphere. It indicates the intensive core kernel activity in every giant planet, and more intensive in Jupiter. . Such phenomena point out that the core kernel in every outer planet is in intensive transformation, still possesses large unconsumed ultradense mass, and does supply the planet with atom-molecular matter.

Why does Mercury have such a large core? - Mercury has a very weak magnetosphere, which is evidence for weak activity of its core kernel. It is why the outer melted part of the core is only a thin layer. Mercury has very large core (occupying 0.7 of the radius), but it is barely half the radius of Earth's core, and occupies barely a seventh of the volume of the latter. Probably, its original core kernel, proportionally with core mass, has been 7 times smaller of the Earth's initial core kernel. Let's see how it was formed[38e]. The very large core of Mercury is explained by its orbital position and core kernel activity. Since its consolidation as a planet, Mercury might have had an orbital eccentricity to the Sun even greater than it has now. During the early phase, Mercury's perihelion probably reached extremely close to the Sun, nearly 4 times closer than that reached by the

Earth's perihelion.

Under such a strong orbital environment, Mercury's dynamic parameters compared with those of the Earth would have had been as follows:

- *Sun's gravitational attraction* => about 16 times greater
- Radiation received from the Sun => about 16 times stronger.
- *Mercury's gravitational attraction* => about 3 times weaker

Consequently, these parameters influenced that the lighter and overheated, outer atom-molecular matter of the still unconsolidated planet might have been attracted by the Sun during the beginnings of the initial (Hadean) phase of the planet. Potentially, in a different environment, the kernel of Mercury could have formed lager mantle and crust, and the core diameter would have to be 0.5 of the planet's one.

The Sun's activity and Earth's core - The view on the outer temperatures of the Sun (fig. 8) shows a certain analogy between the Sun's activity and radiation with Earth's core energetic structure (fig. 1). Such comparison shows that in the Sun and in the Earth's core, similar processes must run, differentiated from environments in stars and planets, but connected by their origin within the same nebula. Further it is need to repeat that seismic study unequivocally determines that earthquake energy comes from the Earth's outer core and that the core activity and earthquake events are influenced by the Sun's cycles. Also, there is an apparent correlation between earth seismicity and sunspot cycles, implicating the Sun's radiation in Earth dynamics[8a,c].

It is observed that Earth's magnetic field is disturbed by the solar wind and also by Earth's seismic activity around the epicenter some hours before an earthquake shock occurs. It is another sign that the core, along with the Sun, also releases plasma. Integrating this evidence with the temperatures, which rise around 170 - 200 times from the

Sun's surface (6000 K) up to the lower corona (> 1 000 000 K), we should conclude that above the photosphere (sun's surface), there are probably continuing processes among particles, bursting outside the photosphere and generated from *a probable ultradense ultrathin* helio-spherical kernel which contributes to making the Sun the way it is, a super giant core with a super giant core kernel.

This leads to an understanding of the peculiarities of the inner core: the solid state of the Earth's inner core just inside the ultradense kernel is conditioned by the negligible proportion of radiation going inwards and by temperatures not being able to be raised above the temperature rate of the lowest part of the mantle at core-mantle boundary. Similarly with the sun's atmosphere, the maximal temperatures ought to be not next to the core kernel, but somewhere in the outer zones of the Earth's core. Finally, these conclusions allow us to interpret the Sun's phenomena through the investigation of the core and vice versa: we observe the Sun to interpret the Earth's core.

8. Earth's Core and Planet-Formation Process

As pointed out previously in this scientific pamphlet, the Earth's core, by its role, is connected directly with geological phenomena and processes, and is, by its origin, linked to star-formation processes. Along integration of the Earth's core data and related evidence from various scientific fields, is not so difficult that we could not see some connections of global tectonics theories with fundamental studies in the quantum field[18], in particle physics[9], and in astrophysics[3],[4]. Consequently, it has come to view that the Earth's core is an energetic cosmic object. Then let us now consider the matter transformation of the core kernel a cosmological problem.

Core in the solar nebula, the potential Earth - As it already is seemed, the independent Earth's core structure, with its energetic source inside, points to its cosmological origin, it's direct indication of the gravitational collapse of the gaseous-dust cloud into the solar disc, as one of multitude of similar cores. According to the current standard theory, our solar system was formed and consolidated by the gravitational collapse of a gas-dust cloud [4.6 billion years ago] of a gas-dust cloud, just occupying a small part of a giant molecular cloud enriched with the remnants of a supernova explosion. From the most of the mass collected in the center of the rotated disc was formed the Sun, while the planets and planet-like objects were formed from the fragmentary rest mass., Here in this current theory is missing the factor triggered the motion of the cloud and respective differenced gravitational collapse in Sun and planets. Furthermore according to astronomers[49], the stars' natal clouds arise as part of a grand cycle inside the medium of intra-galactic space in which gaseous dust material circulates from clouds to stars and back again. Furthermore, the initial phase of a nascent star formation is hidden in denser opaque parts of the clouds. What does trigger the cloud to collapse is still an unsolved problem.

Theoreticians differ on what causes the collapse as to whether it is an external factor such as a supernova explosion which "triggered" the initial collapse of the cloud, or whether it is an innate quality of the cloud. Both these factors are integrated by core kernel theory. Core kernel is part of the supernova remnants inside the nebula and is acting as innate factor. Farthermore is concluded that supernova explosion is the pivotal factor for recycling the cosmic scattered atomic and molecular material from the stars to space and vv from the space to star. Just the core kernel is transitory factor of this grant cycle.

Most recent observations manifest the presence of an infrared source in the center of the cloud, which may be

43

interpreted as an early stage protostar, causing the collapse. This is the first sound idea that imposes a radical change in the theory of cloud-star relations. Thus, the new idea is that an internal object within the gaseous-dust cloud incites gravitational collapse and accretion. Possibly, it could prove to be just the required link between the concept of the Earth's core kernel and star formation. As it's seen, whatever might be the triggering factor, this evidence confirms that the responding properties of such an excitant have the transformable ultradense kernels inside the cosmic cores.

Evaluating these phenomena from the perspective of the role played by an ultradense kernel, the standard theory of Earth's origin would be modified as follows: Processes of planet formation have to begin after a gigantic cosmic outburst of an object like a compact star transformed into a supernova or similar event. Immediately following the explosion, along with radiation, gas and dust, a multitude of ultradense blazing nuclei like granules or beads is probably propelled outward from the center of the explosion at a cosmic speed having even rotating motion. The miniscule ultradense granules followed motion of the largest portion of the ultradense remnant having even their rotating motion. Such burst simultaneously causes the ignition which then causes burning or the transformation process of the *'granules.'* These coalesced to each other, and immediately become mini-radiating objects, the energetic cores of the potential planets, which also eject plasma, radiation and chemical elements inside the gas-dust-debris matter (Fig. 7) [38e]. The new material, mixed with previous material, forms a dense floating gaseous dust cloud around the dynamic granules, and gradually arranges itself into a planetary disc orbiting the Galaxy.

Only such an ultradense matter shape of the initial nuclei-like granules would have the property and tendency to connect by collision or accretion, causing the blazing granules to fuse with each other and form the nuclei (core) of larger objects of the infant planetary system. These gravitational objects gradually attracted and cleaned the space around them from the atom-molecular shape of matter (gaseous-dusty mass). In such a way, the smallest burning balls of the nuclei gradually became planets' cores covered by fiery silicate lava, and further they became proto-planets orbiting the largest one, the potential Sun of the system. The core structure of the infant Earth was developed through the core kernel's transformation and became similar to the structure of the current Earth's core, since the beginning of the development of the sedimentary cover, sedimentoosphere.

This theory supports sound aspects of space mechanics about the Earth's origin in the solar system, but rejects its weaker aspects related to the kinds of initial nuclei or cores to be fragments of the hard grains or debris to be the cause of the transition of a cosmic cloud in solar sytem. Furthermore, this theory is in conformity with the geological history of the Earth's crust.

9. Core Kernel and Formation of the Planets Orbiting Galaxy and Binary Stars

The core kernel, as the essence of the geo-theory of matter transformation, as a new formulation of the Earth's energetic source, could help to impove many other scientific concepts, and to stimulate further progress toward space investigation and exploration. It is in our nature to push forward scientific investigating abstraction in order to seek, discover and explore unknown things, objects and worlds. Let's consider how the core kernel theory explains the independent formation of the planets, through similar way as stars along circulation of the cosmic atter..

Discovery of the amazing phenomena. Astronomers have lastly discovered two amazing cosmic phenomena; planetary systems of double stars, and remote, rogue

45

planets orbiting galaxy independently.

Preliminary view on the Phenomena unknown before - This was the first that theoreticians met a problem that need to find a different way in order to explain how are formed planetary systems of binary stars, and how are formed remote planets.

But very recent observations have found dark mini-clouds, *called globulettes.* This observation supports a hypothesis that probably there are places that might have the right standard conditions to form planets with no parent star required; but surely according to the same way as planets with parent stars are formed. Another interpretation is that such planets were formed with parent stars and were later separated and wandered in galactic space, until some of them entered into gravitational field of the bary center of binary stars. These explanations according to the present standardized theory is an adaption of the same way as previously, when many new and unexpected discoveries were fitted into the same preexisting framework, but now the theoreticians have to reach a new conclusion in order to explain how the planets and stars are formed under an universal law acting in intra-galactic processes.

Energetic core and the origin of the remote planets and the planetary system of a binary star - In brief, both phenomena, the rogue planets and the the planetary system of a binary star verify the core kernel theory. This core kernel theory points to the same process causing the planets' formation and their independent growth similarly to stars. Consequentely, the core kernel mechanism can explain the formation of the rogue planets orbiting galaxy, as well it can explain formation the planets of the binary star system (or multi-stars system). *So, the stars and planets are formed and developed by the same unique process..* The nexus of this theory is derived plainly from the fact that Earth's energy is released from the core, not as something in itself as *'energy from energy'*, but by the reality of *energy* *from matter transformation.* Plainly, if the Earth's energy is released from the core, the core must be an independent cosmic object in a planetary system, and out it. This concept is the pivotal point of the matter transformation inside the core, expressed concisely here in this pamphlet (and previously published at length[38 a-e]. Hence, here in Pamphlet is arguing that the Earth's original core, as one among many similar cores of other planets, was a cosmic energetic object inside the gaseous-dust cloud of a potential solar system, and it was formed by the accretion of identical smaller cores orbiting the gigantic central core, the infant Sun.

In the case of the planetary system of the binary star, the accretion process of the energetic cores has developed as follows: the smaller star, having nine times smaller gravity, behaves as a satellite of the larger one. This means that the minor star, having lager core than a potential planet, instead become a planet, would remained like the satellite star of a large star. When the difference between the cores of the two potential binary stars was extremely large, then the minor star would remain as the burning core of a potential planet. Information from cosmic discoveries confirms that some stars are extremely small. In these extreme cases, the smaller star might be 30 times less massive than its larger twin star and would orbit it as a real burning satellite. or potential planet. In addition to this, some remote, rogue interstellar planets orbiting galaxy, observed recently, are thought to have formed in a similar way to stars, and are proposed that those objects be called sub-brown dwarfs. It is another evidence to confirm that the planets are formed and developed by transformation of theirs core-kernels similarly as sun-like stars and dwarf brown stars. Furthermore, a brown dwarf star was stunning in its transitory state, not attributable to deficiency of the cloud's material accumulation to ignite the nuclear fusion reactions, but its state is due to the size and magnitude of disintegration of its core kernel.. Actually, are approached

47

premises to discover a multi-star disc, where should seen mini-stars orbiting as satellites a super giant star, then the core kernel theory will be the fact.

10. Energetic Core and research Problems

This geotheory of the ultradense core kernel transformation creates a new view on different still disputable theoretical problems in many research fields. It is very important to compile new working hypotheses for research and investigation activities seeking practical results. Let's examine *two main theories about the origin of hydrocarbons* (oil and gases): the biogenic theory and the abiogenic one. *First, the biogenic theory* explains their formation from transformation of the micro-organisms' sediments as *"fossil fuel."* But, it was modified in order to explain the presence of very large hydrocarbon resources in unusual depths; cases of data pointing to the recovery of the explored oil fields and their presence in many cosmic bodies. This modification rejects their origin from surface biota, and suggests that thermophilic bacteria as those of the hot springs of the ocean bottom (at temperatures up to 122 °C, and under the radiation of the chemical reactions), may live in the depth of silicate cover of Earth and the planets. These bacteria allegedly interact in the environment and cause the generation of the hydrocarbons including oil. This hypothesis is evaluated as improbable, as are efforts to link it with the Expansion[36a]. In spite of this, currently, the biogenic theory is still the dominant and standard theory, because the very sophisticated chemical reactions, by which the abiogenic way is reasoned, are currently considered as less likely or less possible to have occurred in those ways. *Second*, the abiogenic theory needs more explanation. It could be explained plainly only by the geotheory of matter transformation: The abiogenic formation of the hydrocarbon molecular bonds most likely occurs as a result of the direct synthesis of the newly formed the atom of hydrogen and carbon through core kernel transformation as other molecular bonds of chemical elements are similarly formed. In an environment, where the oxygen has been exhausted, especially forming silicate minerals and stable oxides including water, the carbon and hydrogen are connecting in theirs hydrocarbon molecular bonds (C_nH_{2n+2}).

11. Earth's Core Kernel, a New Outlook on Cosmos

"Every phenomenon and every existence in every point of the perpetual cosmic space is consequence of matter interaction and transformation" (Author)

As it is seen throughout the article, the Earth's energetic source with the core kernel concept as a cosmic engine is becoming very fundamental to the formation of the planetary system and the evolution of the Earth. It supports a new outlook on permanent matter transformation in limitless space, time, objects, phenomena, processes and in the endless combinations of the atom-molecular shapes of matter. Shortly, every phenomenon and every existence in every point of the perpetual cosmic space is consequence of matter interaction and transformation. So, the aim of this paragraph is to distinguish the concept of the *Earth's growth and development through the core kernel transformation mechanism*[38 a-e] from *Earth's expansion or growth* based in uncertain concepts of the universe-space-time expansion or on the uninterrupted enlargement by the continuing accretion of scattered cosmic matter or of the ether-like space environment[6],[26a],[1].

The interpretation of the core kernel's transformation is even in the accord to the notion of the absolute matter motion in absolute space, and time[16]. It is fact that as a result of matter interaction and transformation in limitless space we can understand motion, and as result of motion we can understand time. Hence, it is necessary to clarify some concepts which universally characterize the matter transformed in the Earth's core, viewing it now as a cosmic energetic body, subject to cosmic notions: *Universe; Space – time curve; Mass – energy; Dimensions.*

Core Kernel and Big Bang universe - The core kernel concept stems from the reality of interaction and transformation. It results from an intra-galactic, big burst (supernova-like explosion) in contrast to the Singular Big Bang aiming impossibility, an occurrence outside the interaction and transformation. By interaction and transformation is functioning the real world, in earth and in universe.

Our process of thinking must properly select whether the images are true or false. In our memory both images may exist, in one way as unreliable tales and myths and, in another way, as the description of well-documented real events based on true evidence. But our reasoning cannot always distinguish the real from the unreal, since both are registered in our memory. It often happens that, in our memory, a theory previously registered as true and well argued and gets inertia and, keeping the inertial state, does reject any contrary theory, even when the new theory is composed even with many new unexpected data and facts. But impressive new facts already enter our memory adapted previously for keeping the inertia state, not changing the inertia of the stabile theoretical course. The key to our understanding is to distinguish what in our memory is recorded as real objective existence and what is placed as unreal image; what is logical and what is illogical to sound reasoning. Evidentially, there is a misunderstanding of the term *"universe."* Furthermore, the term *"universe origin"* brings even greater confusion to our reasoning.

The term "universe" is inherited from Aristotle's

50

geocentric concept and is wrongly reinforced by Ptolemy's calculations. By the origin, the term "universe" has meaning as the entire space around the Earth enclosed by the sky's spherical vault. This concept was modernized and became cosmocentric sky and is conceptually related to the notion of the universe origin. It follows that the term "Universe", regardless of the fact that etymologically it points to the view representing the whole; it cannot represent a sphere that encloses the whole, as it is implied by this notion. Otherwise, such a perception would allude to the Earth as the center of this celestial firmament of Cosmos, newly invented as Universe by the Big Bang singularity of our Era; allegedly the distant galaxies are going away from our observable point with abnormal fast speed. This theory contradicts itself. It is constructed entirely by physical mathematical concepts, interpretations, formulations and calculations, but its starting premise is a point without space, a point without coordinates. Its space coordinates are zero. Mathematically, 0 (zero) coordinates make calculations 0 (zero). Furthermore, scientifically and mathematically, a material point in itself without coordinates and without environmental interaction is meaningless. Even the lastly modified concept, that this singular point was the result of the concentration of the previous universe, is the feeblest support to justify the singularity and the beginning. Bing Bang theory calculates an imaginary vault of heaven of the cosmos. This cosmic sky is in an expanding, rather in inflating process towards gaining space from nothing, from 0 (zero) space, and after does exhaust the expanding process begins opposite process of the return to the zero, to the "death" of universe, and again to the "revelation".

In essence, the singularity of the Big Bang Theory is an imaginary physical-mathematical construction starting from a fantastic premise that the universe began from an initial irrational point. This point, called previously (1927) *primordial atom,* abruptly burst, and became universe, which started to expand, and over billions of years, it is still

51

continuing to expand, as we now know it (Firstly, this hypothetic interpretation has been determined by the founder, a Belgian priest, theologian and cosmologist, Georges Lemaitre 1894 – 1966). So, the Big Bang requires that existence has been created from nonexistence; it is why it was exactly labeled as *'cosmic egg'*, an unfertilized egg, which laid the fertile hen. So, to discuss the big bang scientifically is worthless. Nevertheless, we are obligated to explain that it is baseless theory. The well-known physical law of the universal matter interaction and transformation is, by the big bang singularity, replaced by the creation. So, the big bang singular theory is an attempt to calculate the age of the universe, or to date the creation, similarly as the mythological and biblical hearsay held that the Earth and its entire universe inside the sky, brought into being in *a grand creation event* by the God in certain time, while many began to ponder the question of precisely how long ago this event happened.

Furthermore, the basis of the Big Bang and the expanding universe theory was wrongly taken redshift of spectral light with increasing distance of galaxies, while astrophysicists concluded that it must be a consequence of some other phenomena – something that happens to the light itself as it travels through space.

Many authors are engaged to make Big Bang scientifically more acceptable, and they have coned many sub-theories under the notions: "the observable universe", "our universe", "other universes" "singular multidimensional universe", etc. These big bang's sub-theories allude for many isolated singular points as potential universes, out of the permanent interaction. In this way, such reasoning makes a vicious circle in order to support the singular point of the Big Bang, as an effort to approach to the scientific notion of limitless space of unlimited ways of the interaction and transformation, similarly to: plus - minus, acid – base, particle antiparticle. Furthermore, some last interpretations[21] are pursuing this trend, a certain cosmic radiation was newly caught by Antarctica's cosmic observatory. This radiation arbitrarily was labeled as graviton flux from the ultra-moment of the big bang, of the singular point explosion, and it was considered that the moment of the creation was fixed.

The notion of "our universe" lastly is theoretically modified[2] allegedly so: In higher dimensional universe a black hole singularity has burst as an ultra-supernova explosion, and it created our three dimensional cosmos. It alludes to our observable three dimensional universes inside the multidimensional universe of Big Bang. This theorization is meaningless as well. Factually, illusive term "our universe" is an observed space, a portion of spherical space with a radius progressively enlarged simultaneously with the improving technology of the observation. Why is such an expression meaningless? Because it is fact, the space does not end precisely where our observation ends; beyond observable sphere of space is spread limitless space.

Let us imagine observers installed in the multitude of the peripheral galaxies of the observed universe (space). Our galaxy is in periphery of their observable space. Around each of them is extended the space, in the same way as for us, i.e. in this part and beyond the limit of our observed space, with the size of the radius that both sides look at each other. Thus, moving the image from our **observed space**, to the analogues observable peripheral space and further on continually, we can perceive the infinity of space where matter is cyclically transformed from a very dense state into a very rare one and vice versa thus producing the common atom molecular shape of matter. This transformation occurs in the Galaxy but, compared to this Galaxy, on a ultra-small scale it occurs within the Sun and compared with the last one it is occurring in miniature within the core of the Earth. While the Earth core originated as a ultradense remnant of a supernova explosion.

Space-time curve - The existence of Matter as mass, as

scalar quantity, is naturally perceived with its required space, while, time, as vector quantity, is perceived as a consequence of matter motion, by which the age of events is measured. The time is perceived by its measuring unit of a relative motion. If wave communication among closed systems is done in a curved path, this doesn't mean that time depends on the curvature of this path. It is the same as saying: seeing that the trajectory of a projectile is curved, the distance from the gun to the objective is the bent line, crossed by the projectile and not the real straight distance between both points. Furthermore, it is meaningless to judge that curvature of a path influences respective time to bend too. Every movement in cosmic space of objects, radiation, motion of the particles and sub-particles is influenced by interactions, and cannot be straight line, it does form, of course, excessively complicated movable systems, and the final path of its movement is always a curve. Also, it is clear that, energetic matter fluxes, moving by curved path toward ultra-compact objects, does not cause space-bending or time-bending. Only a curved surface can enclose a space of certain dynamic objects or different systems arranged in chains in unlimited space. The so called "*observable universe*" is a spherical fragment of the space reached by observation or by calculation; it is full of groups or a super-groups of galaxies. Then, in the reached space of observation and beyond it, the bending of the path of light and the path of every radiation is not a bending of space-time, even through the event horizon when, as it is supposed, the matter density is going to the infinite[4].

The Time measuring unit, as a parameter of motion, is determined by Earth's rotation around itself, the day, and around the Sun, the year. To say space-time bending seems the same as saying distance-year bending, concretely *centimeter-second bending.*

Energy-matter – Sometimes, efforts are made to disconnect energy from matter, respectively from mass. This idea is based on the disconnecting concept of the particles

54

enclosed in an atom's structure from the particles and sub-particles, free and in motion, that form wave-corpuscular matter and wave fluxes of the radiation. Every particle, inside or the outside of the atom structure, has mass. It means that energy is not something else, but is a matter shape of particle fluxes in motion as a result of the interaction. If a common object is accelerated towards the speed of light, then the atom-molecular object must be disintegrated and transformed to a different form of energetic wave corpuscular matter. The measuring unit of energy is the electron volt (the mass equivalent of 1 eV is 1.783×10^{-36} kg). It demonstrates that energy is matter or rather matter in motion. Mass-energy relation is expressed mathematically by Einstein's formula $E=mc^2$. That means that energy is mass, the law of matter transformation.

Dimensions - Frequently, one hears about the fourth or fifth dimension and also multidimensional concepts of curved time-spaces. In this way mathematical parameters are confused with space coordinates. Space is permanently a Euclidian space of three dimensions. Surely, space is perceived by the presence of the matter in different objects and wave-corpuscular interaction among them. The motion of matter determines time which is expressed as a vector quantity. The calculating parameters might vary, depending on the properties of the relevant surface enclosing the certain space, and on the hierarchy of the motion. For example, a cube or a sphere is three-dimensional, but to figure out its volume each uses only one parameter, the side or the radius. If we should calculate the volume of a cosmic space, closed by complicated movable curves, we should use many parameters according to the enclosing curved surfaces. The counting parameters are not space dimensions or space sizes. So, the speed of motion and the motion as measure of the time are parameters, not dimensions.

Concluding Remarks

Judging about current theories of continental shift, we are far away from the situation of the century before, when the hypotheses of continental shift were founded. Actually, we have unimagined facts arguing the reality that continents originally were next to each other divided by global cracks, and because of the growing earth process the continental blocks moved away, while the cracks has been open, and the space between crack's parturitions was immediately fulfilled with new basaltic crust. So, the continents departed away from each other as parts of the lithosphere slabs (Plates) together with respective oceanic crust.

These unusual achievements were new premises to review the current theories and to uproot hidden archaism inside them. Then, let's try to recap the essential points of the theory, and express concisely what it means to shift earth energy source from the silicate cover of the core into the core interiors.

> ➢ The Earth's is an energetic cosmic object inside its inert rocky silicate shell. It is why the expansion becomes Earth growth from matter transformation of the kernel of the core positioned, as an ultradense diaphragm-like screen, somewhere between both cores. By this spontaneous disintegration, the particles and ultra-particles are released from their ultradense state, and, on one hand are reorganized as radiating fluxes of the physical fields and every radiation, and on the other hand do compose the nuclides of plasma-like mass and are further reconstructed in atom molecular bonds, in unusual greater volume than previously in ultradense state, as additional mass to the growing earth process.

56

> Simultaneously with ocean bottom growth, the entire Earth is growing especially the entire silicate cover and the inner core, while outer core is in dynamic equilibrium or is growing slightly. Furthermore, the radius of the core kernel is enlarged simultaneously with its disintegration.

> The Earth's core with its core kernel was the embryonic Earth inside the planetary nebula, while the Sun was formed as unusually larger central core of the infant planetary system. This specific planetary cloud was probably formed through reorganization of the ultradense energetic remnants of a huge intra-galactic explosion like a supernova. The similar transformation process occurs even for formation of the planetary system of binary stars, brown dwarf stars and remote, rogue interstellar planets orbiting galaxy; similarly as sub-brown dwarfs, are formed that way.

> The core kernel, as the mechanism of the growing earth, explains more properly and in a new way geotectonic problems; all Earth's phenomena and processes, especially the problematic phenomena such as the complete earth inner structure, the presence of continental rocks inside the oceanic bottom; the heterogeneity of the mantle and the relation of the mantle plumes to core-mantle transition D" layer, etc.

> The difference in the state, activity and in the core size of the planetary bodies depends on the current state and the intensity of the core kernel transformation.

> Acceptance of this new concept as a working hypothesis would offer epochal advances in the fundamental sciences, especially in physics and

general knowledge, and would influence the research orientation concerning both the Earth's environment and cosmic space. Furthermore, Earth's core must be seen as a star-like Sun in miniature and its investigation urges us to understand the Sun better, while observation of the Sun helps us to understand the Earth's core function better.

*   *   *

E p i l o g u e

Earth's Core Kernel

Already, it has become a well-known geological fact that radioactive isotopes and their last products of disintegration, in the entire silicate shell of the Earth's core, are in very unimportant amounts. This reality demonstrates clearly that the respective radioactive disintegration could not produce sufficient energy to be the perpetual source of geodynamic phenomena and processes. Hence, the attention of the geo-researchers to find another source and another way of energy generation is naturally orientated toward earth core. But, the stabilized concepts, imaging the core as a heavier precipitate from newly melted silicate mass of juvenile earth, excludes the core from being the source of the requested energy. Then, do we, the geologists and other geoscientists, be silent, even when we know that the progressively published new data already diverge from the older interpretation of energetic source, even, when it is standardized *under the Big Bang* of the unmerited mathematical prestige? Surely, we the theoretically advanced researchers in geology are researchers of one best known planet, and unwillingly did become planetologs and may consider ourselves capable to integrate the achievements in astrophysics with those of the geophysics, and   find a new solution that leads us to unite both the rock-formation process (our field of research) and the energy release (the cause of rock-formation and deformation) as a consequence of the same universal process.

This reality naturally would lead us to the necessary conclusion that the energetic source must be in the shape of a hearth, a focus inside the core. This source must be from a certain ultradense matter shape in spontaneous disintegration of particles and ultra-particle, which on one side are reconstructed in atom-molecular matter and occupy unusually greater volume than that in the previous ultradense state, and so causes the growth of the entire Earth; on another side they cause the physical fields and different radiations. By such function, the energetic source becomes *the kernel of the core*, shortly a *core kernel*. Conclusively, only proper and natural position to place the core kernel is the boundary belt between outer melted core and inner rigid core. There between both cores, the core kernel must be in a form of a spherical diaphragm-like screen in ultradense, ultrathin and in indiscernible state.

The energetic kernel of theEarth's core is confirming grand matter circulation from the compact ultradense massive stars to supernova phenomenon (explosion) and to the space radiation, space nycleosynthesis , and further back.

So, core kernel function dictates new interpretation of the planet's cores origin since solar system formation. Probably, inside the potential planetary cloud was a multitude of the energetic ultradense droplets or beads as remnants of huge intra-galactic explosion. Just, these remnants have to cause transformation of dusty cloud in solar system. Firstly, to every planet was by accretion formed planet's energetic core (kernel with crown), and afterwards was formed rocky silicate covering by core kernel tgransformation and by attraction of the space material of the planetary cloud. So, it seems that planets are developed independently inside or outside the solar systems starting as stars in miniature.

Consequently, due to the integrated facts pointing, with highest probability, to the core kernel theory, the Earth's core, sooner or later, should be evaluated as an energetic cosmic body by theoreticians and other thinkers of the Earth and the Cosmos.

In addition, the wider opinion will be interested to know that my theory was encouraged and endorsed by following scientific authorities:

- *Professor Cliff Ollier* of Australia wrote a preface of this theory previously (2005), and in this pamphlet, as well as acting as scientific editor.

Professor Giancarlo Scalera encouraged my participation previously in Urbino Workshop (Aug. 29-31. Italy) and in Erice Conference (Oct. 4-9 2011, Italy), and lastly chose and published my paper in the Selected Contribution of Conference (Edition of. National Institute of Geophysics and Volcanology, Roma 2012).

- *Professor Salvatore Bushati* of the Academy of Science of Albania has evaluated the theory as a basis to put forward the leading thought about Earth and Cosmos.

- *Professor Adil Neziraj*, General Director of Geological investigation of Albania, and *Professor Afat Serjani*, President of Geo-monuments' Association of Albania organized a scientific conversation (photo below) and did the respective publication of this theory.

Furthermore, I also feel obliged to inform the reader when was the initial moment I coincidently determined the concept of the relation of the "matter-energy unity" with the Earth's energetic source. It was when *Professor Alfred Frasheri* of geophysics and his team invited me in a workshop with following topic *"The Earth's physical fields, a manifestation of the matter"*. The next day, an article about this event, included the following phrase that expressed the essence of my discussion: "The geologist Vedat Shehu showed that . . . . . . , the removals of the galaxies may indicate expansion of metagalaxy, but never of the universe that is infinite and permanent." (Student Newspaper, April 23, 1970. Library of Tirana University). Three years later, I started the systematic study to prove that the cause of the geodynamic phenomena and processes is interconnected to the cause of the solar system formation.

Some evaluations

Vedat Shehu has produced a remarkable book that I commend to all Earth scientists.....an unusual viewpoint, which should provide a breath of fresh air in many geological debates."

Professor Cliff Ollier
Neflands, W. A. 6009, (08) 9380 2664. [From the preface of the book, USA 2005, and edited in NCGT Newspaper No. 38, 2006)

"I would like to consider Vedat Shehu the evolutionist of today's science. His interpretations I hope will become a basis of discussions in order to put forward the leading thought about explanation of the universal processes which occur in Earth and in the cosmos."

Professor Salvatore Bushati.
Academy of Science of Albania, Geophysicist, (in evaluations of the book "The Growing and Developing Earth.", in Albanian 2009)

"... Your's way of scientific integration attracted my reasoning, and as I catch, .., probably it might attract attention, and encourage others to pursue proof of existence of a certain core's energy source as a mechanism of the Growing/Expanding Earth Theory."

L. Joan Brown.
English language editor, Sharon MA Public Library, Literacy co-operation for Volunteers of MA. Jan. 2011 (From the correspondence of the preparation the Pamphlet in literary scientific English).

Acknowledgments

Grateful thanks and acknowledgments to my colleagues helping in improving and integrating it into the compact form used in some articles.

Special thanks to *Professor Clifford Ollier* of the University of Western Australia for his suggestions for every one of my editions and for the prefaces to some of my works including this Scientific Pamphlet, and to Professor *Giancarlo Scalera* of the Institute of Geophysics and Volcanology of Roma for stimulating the specificity of my theory within the Expanding Earth Hypotheses, and for some related suggestions.

Special remark: I am grateful to the language editor of this Pamphlet, *Mrs. L. Joan Brown who,* having also read key parts of *The Growing and Developing Earth,* gave me enormous assistance to realize the reconstruction of my theory as concisely as it is expressed in this Scientific Pamphlet. Mrs. Brown, , proposed and encouraged me to write the essentials of the cause of the Earth's dynamic processes in a more understandable and simpler way. *Mrs.Brown* has previously assisted me with the writing of the scientific paper, which I presented at the Erice Conference in Sicily. This paper was, one between others, chosen for publication after the conference, and It was placed in the Section of *Physics & Cosmology*, in the *Selected Contribution of Conference*[34].

Vedat Shehu
Email: vedshehu@yahoo.com
115 Norwood Main St. Sharon, MA, 02067

REFERENCES

1. Adams, N. (2006). "New Model of the Universe." <http://www.nealadams.com/nmu.html 27 Adams 27.

2. Afshordi, N. Mann, Robert, B. and Pourhasan, R. (2014). The Black Hole at the Beginning of the Time. Scientific American. 311 (2) 38-43.

3 Anderson, S. F., et al. (1999). "Mapping low density galactic gas: third helium Lyman-alpha forest." Astronomic . 117, 56-62. DOI: 10.1086/300698; e-print: astro-ph/9808105 | PDF.

4. Barcelo, C., Liberati, S., Sonego, S., Visser, M. (2009). "Black Stars, Not Holes." Scientific American 301 February 46-52

5. Berhe, S. M. (1999.) "Ophiolites in Northeast and East Africa: implications for proterozoic crustal growth." (London: Journal of the London Geological Society; V. 147; No. 1, 51-57.

6a. Blinov, V. F. (1973). "On the hypothesis of Earth's expansion." (In Russian). Fizika Zemli 1 27-35.

6b. Blinov, V.F. (2012) "Geophysical Advances in Earth's Evolution – Kinetic Gravity and Expanding Earth. " 173-184. In "The Earth Expansion Evidence, A challenge for geology, geophysics and astronomy" Selected Contribution to the Interdisciplinary Workshop, held in Erice, Sicily, Italy (4-9 October 2011). . Post- conference publication edited by Giancarlo Scalera (editor in chief), Enzo Boschi, and Stefan Cwojdziński. Rome,2012. <http://www.aracneeditrice.it/pdf/9788854856936.pdf>

7a. Carey, S. W. (1988). The Expanding Earth Amsterdam (Elsevier Scientific Pub. Co., 1976). 488

7b. Carey, S. W. (1978). Theories of the earth and universe: a history of dogma in the earth sciences. (Stanford CA: Stanford University Press, 1988). 436.

8a. Choi, Dong R. (2010). " The January 2010 Haiti Seismic Disaster Viewed from the Perspective of the Energy Transmigration Concept and Block Tectonics." NCGT

Newletter, 54,. 36-54.

8b. Choi, Doug R. and L. Maslov. (2010) "Global seismic synchronicity. " NCGT Newletter, 55, 66-74.

8c. Choi, Doug R. and L. Maslov. (2010). "Earthquakes and solar activity cycles." NCGT Newletter, 57 85-97

9. Close, Frank. (2004) "Particle Physics, a very short introduction" (Oxford: Oxford University Press.. 160. ISBN 0-19 280434-0.

10. Condie, Kent C. (1997) "Plate tetonics and crustal evolution." Fourth Edition. (Oxford: Butterworth-Heinneman, An Imprint of Elsevier Science Linacre House, Jordan Hill, Oxford OX2 BDP 200 and Wheeler Road, Burlington, MA, USA. 282

11. Cwojdzinski, G. (2012) "Geological Evolution of the Sudety Mts. (Central Europe) on the Expanding Globe." In "The Earth Expansion Evidence, A challenge for geology, geophysics and astronomy. Selected Contribution to the Workshop, held in Erice, Sicily, Italy (4-9 October 2011). 263-273. Post-conference publication edited by Giacarlo Scalera (editor in chief), Enzo Boschi, and Stefan Cwojdziński. Rome, 492

12. de Hilster, D. (2008) "The Growing Earth." <www.dehister.com/docs/TheGrowingEarth.ppt>, 77.

13. Dimitriev, L. V., Vinogradov, A. P., and G.B. Udentsev. (1971) "Petrology of ultrabasic rocks from rift zones of The Mid-Indian Ocean Ridge." *Philosophical Transactions of the Royal Society of London. Series A Mathematical and* Physical Sciences, V. 268, No. 1192. A discussion on Petrology of igneous and Metamorfic rocks from the Oceanic Flore. (London: The Royal Society,). 403-408. Article Stable URL:

14a. Dilek ,Y . et. al. (2003). " Development ideas of the origin of Albanian ophiolotes." In , Y. and S. Newcomb (eds. Concept and Evolution of Geological Thought: Geological Society of America. Spetial Paper, 351 – 363.

14b. Dilek, Y., Shallo M. and H. Furnes. (2005) "Rift-drift, seafloor spreading and subduction tectonics of Albanian ophiolites." International Geology Review V. 47. (New York:

Taylor & Francis Group. 147-176.

15. Dobretsov, N. L, Konnikov E. G., and N.N. Dobretsov. (1992.) "Precambrian ophiolite belts of southern Siberia, Russia, and their metallogeny." In: Precambrian Research 58. In Russian. (Cambridge, MA, Ellsevier. 427- 446.

16. Erickson, F.P. (2008). Absolute space, absolute time and absolute motion. 267.

17. Hilgenberg, O. C.(1933/2003) "The Formation and development of Earth: contraction or expansion." In: Why Expanding Earth? (Eds) Scalera, G., Jacob, K. Proceedings of the Lautenthal Colloquium held on May 26, 2001 in honor of Ott Christoph Hilgenberg. Rome (2003).

18. Hooft, G. (2007). "The conceptual basis of quantum field theory." In: the Oxford Handbook of Philosophy and physics. (Ed. Robert Batterman. 661-729.

19. Hoshino, M. H. (1998). *The Expanding Earth evidence: causes and effects. Tokai, Japan:* University Press, 295. <http://www.dino.or.jp/hoshino/expanding.html>

20. Kolvankar, G. V., More, S., and N. Thakur. "Earth Tides and Earthquakes." NCGT Newsletter, 57. (2011). 54-8

21. Krouss L. (2014). A Beacon from the Big Bang. Scientific American 4, 311. 59-67.

22. Low, F. S. and Kristna, S. (1970).:" Narrow bond infrared photometry of alfa -tory." Nature: 3. 23. 13-22

23. Makarenko G.F. (1983). "Volcanic Seas on Earth and Moon." (In Rusian), (Moscow, Izdatel'stvo Nedra,,

24a Maxlow, J. (2005) Terra Non Firma Earth: Plate techtonics is a myth. Electronic ed. ISBN 0 952 2603.. 189.

24b. Maxlow, J. (2012) "Global Expansion Tectonics: Definitive Proof." 41-60. " in "The Earth Expansion Evidence, A challenge for geology, geophysics and astronomy" - Selected Contribution to The Interdisciplinary Workshop held in Erice, Sicily, Italy (4-9 October 2011). . Post-conference publication Edited by Giancarlo Scalera (editor in chief), Enzo Boschi, and Stefan Cwojdziński. Rome, 492.

<http://www.aracneeditrice.it/pdf/9788854856936.pdf>

25. Moores, Eldridge M., Kellogg, Louise H., and Yildirim Dilek. (2000). "Tethyan Ophiolites, mantle convection And tectonic historical contingency: A resolution of the ophiolite conundrum." GSA., Inc., Special Paper #349 in Ophiolites and Oceanic Crust: New Insight from the Field Studies and the Drilling Program, 349, 3-12.

26a. Myers, L. S. "Earth expanding rapidly by external accretion and core expansion. In:" Global Tectonics Urbino, Italy Workshop: Aug. 29-31. (2004

26b. Myers, L.S. (2008) "A growing and expanding Earth is no longer questionable." (Washington, D.C.: American Geophysical Union, Spring Meeting,. 26a. Myers

27. NASA's Kepler Discovers, "Multiple Planets Orbiting a Pair of Stars", Aug. 28, 2012: *Earth. At:*
<htp://nasasearch.nasa.gov/search?utf8=%E2%9C%93&aff iliante=nasa&query=KEPLER-47&commmit=Search>

28. Noel, D. (1989).Fixed-Earth and Expanding-Earth Theories - Time for a Paradigm Shift? -Version 2, : http://www.aoi.com.au/bcw/FixedorExpandingEarth.htm

29a. Ollier C..D. (2003). "The Origin of Mountains on Expanding Earth, and other hypotheses", in vol. "Why Expanding Earth? "(ed.) Scalera G. : and Jacob K-H. Rome (2003).

29b. Ollier C..D. (2006). "A plate tectonics failure: the tectonic cycles and conservation of continents and oceans." Annals of Gephysics. Supliment to v. 49, No.1, Chapter 8, 27-436.

30. Orlenok (1983)., V., et al. "Global volcanism and oceanization of the Earth and planets." NCGT Journal, V.1, No.1 (Australia: NGCT.org, (1983).

31. Owen, Hugh. "The Earth Is Expanding and We Don't Know Why." in New Scientist, No. 22, Nov. 22, 1984. 27-

32. Puchkov, V. N. "The evolution of the Uralian orogen." (London: Geological Society, Special publication, V. 327, 2009), 161-195. DOI: 10.1144/SP327.9.

33. Romanowicz, B. and Y. Gung. "Superplumes from

the Core-Mantle Boundary to the Lithosphere: Implications for Heat Flux." Science 96.5567. (Stanford, CA: Highwire Press, 2002). 513-516. DOI: 10.1126/science.1069404.

34. Scalera Giacarlo (2011). "The Earth Expansion Evidence, A challenge for geology, geophysics and astronomy" . Contribution to the Interdisciplinary Workshop, held in Erice, Sicily, Italy (4-9 October 2011). . Post-conference publication edited by Giancarlo Scalera (editor in chief), Enzo Boschi, and Stefan Cwojdziński. Rome (2012), 492.

35a. Scalera, Giancarlo (2003) . 'The Expanding Earth: sound idea for the new millennium." In Giancarlo Scalera, and Karl-Heinz Jacob (eds): Why Expanding Earth? :. *Proceedings of the Lautenthal Colloquium, held on May 26, 2001 Honour off Ott Chistoph Hilgenberg.* INGV, Rome.

35b. Scalera, Giacarlo (editor in chief): Hilgenberg, O. C. (2003/1933/ 1939) "Formation and development of the: contraction or expansion." In Giancarlo Scalera, and Karl-Heinz Jacob (eds): Why Expanding Earth? : *Proceedings of the Lautenthal Colloquium, held on May 26, 2001 Honour off Ott Chistoph Hilgenberg.* INGV, Rome 2003

35c. Scalera, Giancarlo, Boschi, E. and G. Cwojdzinski (2012). "The Earth Expansion Evidence, A challenge for geology, geophysics and astronomy. Selected Contribution to the Workshop, held in Erice, Sicily, Italy (4-9 October 2011). . Post-conference publication edited by Giacarlo Scalera (editor in chief), Enzo Boschi, and Stefan Cwojdziński. Rome, 492.
<http://www.aracneeditrice.it pdf/9788854856936.pdf>

35d. Scalera, Giancarlo. (2011). "South American volcanoes and great earthquakes." Article held in Workshop: "The Earth Expansion Evidence, A challenge for geology, geophysics and astronomy - Erice, Sicily, Italy (4- -9 October. 2011). Giacarlo Scalera (editor in chief), Enzo Boschi, and Stefan Cwojdziński. Rome, 492

35e. Scalera, G. (2006). "The Mediterranean as a slowly nascent ocean." Annales of Geophysics, Supplement to V.49,

No. 1, 13-21.

36. Scalera, G. (2009). "Roberto Montovan (/ 1889/1899). In: Biography of a forerunner of the expansion of the ocean floor and the Earth." Montovani and his ideas on the expanding Earth, as revealed by his correspondence and manuscripts. Annales of Geophysics, 52(6), 615-648.

36a. Scalera, Giancarlo. (2012). "Bionic/Abionic hydrocarbons origin – Possible role oof tectonically active belts." In: The Earth Expansion Evidence, A challenge for geology, geophysics and astronomy. Selected Contribution to the Workshop, held in Erice, Sicily, Italy (4-9 October 2011). Post-conference publication edited by Giacarlo Scalera (editor in chief), Enzo Boschi, and Stefan Cwojdziński. Rome, 463-475.

36b Scalera, Giancarlo. (2012). "The volcanic-sseismic clock of the Suth American Pacific margin – A Possible first link between natural disaster prevention and Expanding Earth. In: The Earth.Expansion Evidence by Giancarlo Scalera, Enzo Boschi, and Stefan Cwojdzinski. (Erice, Sicily. "Selected Contributions to the Interdisciplinary Workshop," 4-9 October. 2011). Rome, 479-492.

36c. Scalera, Giancarlo. (2012). "If space is material, what inertia should be? – Rediscovered a dismissed awareness of Ernest Mach. In: The Earth Expansion vidence by Giancarlo Scalera, Enzo Boschi, and Stefan Cwojdzinski. (Erice, Sicily. Selected Contributions to the Interdisciplinary Workshop," 4-9 October. 2011). Rome, 239-242.

37. Shannon, M. C. & Agee, C. B. (1998). Percolation of core melts at lower mantle conditions. *Science* **280**, 1059–1061.

38a. Shehu, V. (1988). Developing Earth (In Albanian). Tirana, Albania. (Sht. Bot. 8 Nëntori, 180.

38b. Shehu, V. (2004). "The Earth: a sample of universe in our hands, according to the Earth expansion through growing and developing processes," New Concepts in Global Tectonics. (Urbino (Italy: Workshop, Aug. 29- 31,

38c. Shehu, V. (2005). "The Growing and Developing Earth." (No. Charleston, S.C.: BookSuege, LLC (2005), ISBN 1-4196-1963-3, USA, 218.

38d. Shehu, V. (2009) The Growing and Developing Earth (In Albanian). Tiranë, Albania: Sht. Bot. Dudaj,. 361.

38e. Shehu, V. (2012). "Earth Expansion through Activity of the Earth Core-Kernel as an active cosmic Object." In: The Earth Expansion Evidence, A challenge for geology, geophysics and astronomy. "Selected Contributions to the Interdisciplinary Workshop," (held in Erice, Sicily, Italy 4-9 October. 2011). 243-262. Post-conference publication edited by Giacarlo Scalera (editor in chief), Enzo Boschi, and Stefan Cwojdziński. 263-273. Rome.

39. Shen, W. B, et al. (2008). "The expanding Earth: evidences from temporary gravity fields and space geodesic GEPH. Research Abstracts V. 10 EGU2008-A-0473.

40. Smith, A.G. (2006). "Tethyan Ophiolite emplacement, Africa to Europe motion, and Atlantic spreading." In "The Tectonic Development of the Eastern Mediterranean Region" A.H.F. Robertson and D. Mountrakis, (Eds.). (London Geographical Society, Special Publication 260, 1-9.

41. Storetvedt, K. M. (2010). "Falling plate tectonics–rising new paradigm: salient historical facts and current tuation." NCGT Newletter, 55, 4-34.

41a. Storetvedt, K. M. (2014). "When Global Tectonics became a 'Pathological Science' ." Essay in; NCGT Journal, vo 2, no4, 106-121.

42. Vogel, K. "Contribution to the Question of Earth Expansion Based on Global Models." (2012). In: The Earth Expansion Evidence, A challenge for geology, geophysics and astronomy. "Selected Contributions to the Interdisciplinary Workshop," (held in Erice, Sicily, Italy 4-9 October. 2011). Post-conference publication edited by Giacarlo Scalera (editor in chief), Enzo Boschi, and Stefan Cwojdziński. 161-172. Rome.

43. Wegener, Alfred. "The Origins of continents and oceans." (Dover Earth Science: 1915). Originally presented at A yearly meeting of the German Geological Society (6 January, 1912).

44. Welsh, W.E. and L.R. Doyle. "World with two stars," Scientific American 309 (5): 4. (Nov. 2013). 40-47. DOI: 10.l038/scientificamerican 1113-40.

45. Wood J. A. "Meteorites and the origin of planets." (New York: The McGraw Hill Companies, 1968) 117.

46. Yano, T., Vasiliev, B.I., Choi, D.R.et al. " Continental rocks in Indian Ocean." NCGT Newsletter 58, (Australia NGCT.org, 2011). 09-28

47. Youcheng C., et al. (1998). "A new interpretation of the Himalayan orogenic belt." Chinese Science Bulletin, 43.1, 83-84. DOI: 10.1007/BF02885523

48. Young, C. J., and T. Lay. "The core-mantle boundary." Earth Planet Science Annual Review, 15, (1987). 25-46.

49. Young, T. Erick. "Cloudy with a chance of stars." Scientific American V. 302. (2010). 34-41. DOI:10.1038/scientific american 0210-34

50. Zagorevski, A. et al. "Tectonic architecture of an arc-arc collision zone, Newfoundland Appalachians." Annals of Geophysics, Supplement to V.49, No. 1., Special Paper #436 in Draut A., Clift, P.D. and D.W. Scholl (Eds.). "Formation and application of the sedimentary record in arc collision zones." (Boulder, CO: Geographical Society of America, Inc., Special Paper #346, 2008). 309-334.

51. Zolensky, M. E. et al. "Mineralogy and petrology of Comet 81 P/Wild 2 Nucleus Samples." In Science, V. 314, No. 5806. (Stanford, CA: Highwire Press, 2006). 1735-1739.

The Picture, along Conversation on the Theory

The picture. In picture along conversation on the Theory (Tirana 25 July 2011) : From left to right: Professor Afat Serjani, the organizer of the conversation and presentation of "The Growing and Developing Earth" theory; the Author Vedat Shehu; Professor Adil Neziraj, General Director of Geological Investigation of Albania.).

E I G H T   F I G U R E S

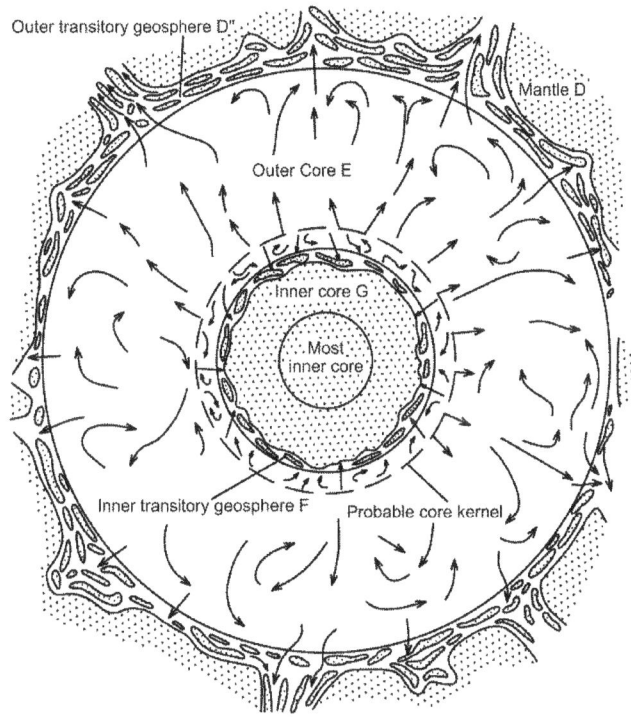

Figure 1. Dynamic Structure of the Earth's Core
The Core's varied concentric spherical structure points to the source of Earth's dinamism, the core kernel being positioned somewhere at the contacting belt of both the inner and outer cores. The Core's spherical layers are identified by the letters from the center outward. The letters represent the following: G, the solid inner core (dots) and its innermost core (bold dots); F, inner transitory layer/geosphere (blank with flakes); Probable indiscernible core kernel (circular interrupted line); E, outer melted core (bent arrows outward and inward of core kernel); D", outer transitory layer/geosphere (blank with flakes); D, the solid mantle i.e., core's cover (dots). The growth occurs inward and outward of the core kernel. (Author)

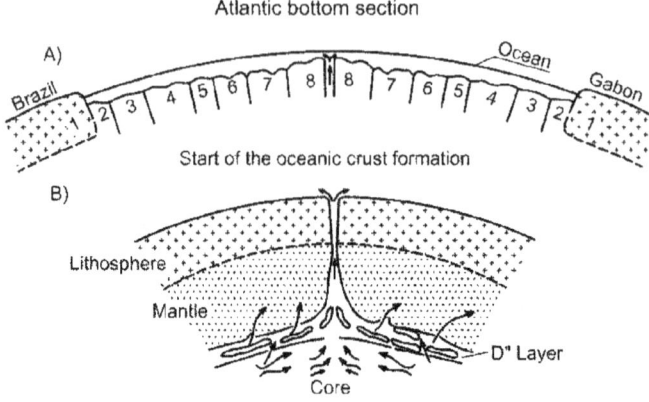

Figure 2. A schematic profile across the Atlantic Ocean
bottom points to growth of the ocean crust as evidence of the
Earth's growth.

A. Current continental slices (1), being previously in contact
(at smaller globe (B) , has successively gone apart from the mid-
oceanic rift, and the space left behind was filled simultaneously
with newly generated oceanic crust forming the consecutive belts
(2 up to 8) of oceanic crust symmetrically shifted from the rift. It
means that the oceanic crust generation is evidence of the
growing earth process.

B. Initial moment of the split of the the continental crust at
small Earth, and Atlantic crust formation, the obvious growing
process caused by core kernel transformation positioned as
ultrathin ultradense geosphere between outer melted core and
inner solid core (fig. 6. (Author)

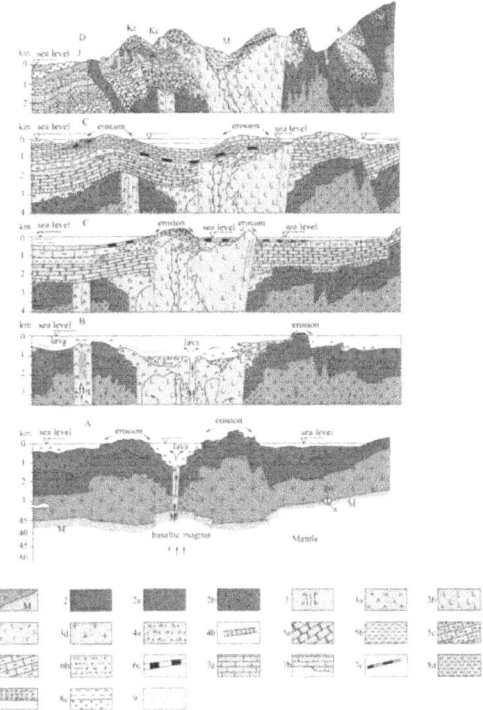

Figure 3. Schematic display of the structural development of the Albanian ophiolite belt:
 Above greenish-blue ophiolites (3, 4) between red and dark violet platform blocks (1, 2) through the growing process has been developed the Alpine cycle (5– 8 in green) in Albanides; moments:
A - In the dawn of Alpine tectonic cycle first transgression on eroded surface (Verruccano's formation).;
B - Early – The beginnings of the Middle Triassic ophiolite belt formation;
C – After Triassic-Jurassic development, the Early Cretaceous regression occurred, when the ophiolite zone (M) was upraised eroded and then occurred the transgressive overlap of conglomerates, laterites, and next of carbonate deposits of the Cretaceous age;
C2 - Paleocene, when third erosion and transgression occurs;
D - Current epoch. (Author)

75

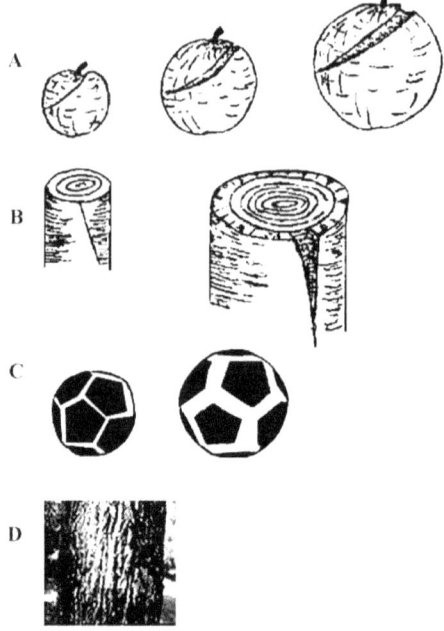

Figure 4. The notions: *the Earth's expansion* **and** *the Earth's growth and development*

The expanding process doesn't occur by matter transformation inside the enlarging object (C), it looks more as inflation. On the contrary, the growing process results from matter interaction and transformation as a universal law of everything. (A, B, D)..

A. The opening of the cutting of an orange through the growing time forms a furrow exposing the pulp between the   rind's partitions.

B. Initial moment of the trunk bark cutting, and the later moment, through growing process of the tree.

C. The moving apart of the pieces on a ball after it has been inflated.

D. Furrow formed in the dead tree bark through trunk's growing process. (Author)

INNER GEODINAMIC PHENOMENA

Figure 5, Schematic block-diagram of the inner geodynamic phenomena and processes
The last part of magmatic mass, generated by core kernel transformation, is flown outward, forms plume structures, causes the mantle growth, the oceanic crust growth, and through inherited ways, it intrudes into the lithosphere and the crust, where might create gravitate disequilibrium and, by replacing it, earthquake shock, or slow tectonic displacement occurs. Its last portion attains the surface as volcanic lava, which immediately fills resulting gap which, because of the Earth's growing process, would be created by the moving away of the lithosphere slices (plates). The process of the gravitational equilibrium replacement, is the cause of the global movable belt formation, stretching force, stress folding and stress thrusting. (Author).

77

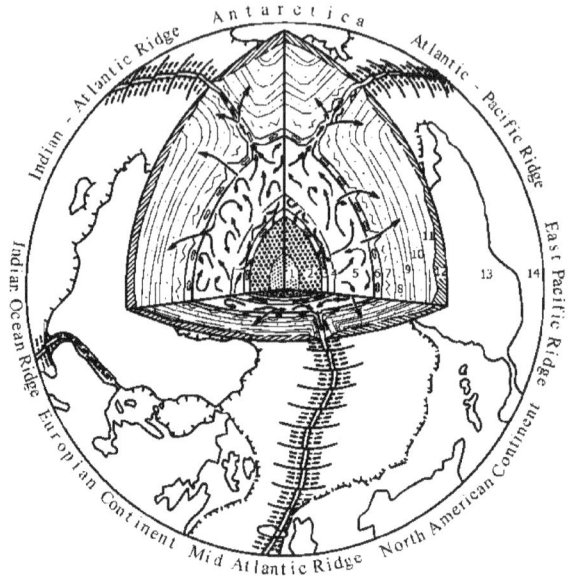

Figure 6.An idealized scheme of the Earth globe growing outward and inward from the transforming ultradense *core kernel (4).*

The growth occurs inward and outward of the core kernel (4): *Outward growth* – previously newly generated atom-molecular and energetic matter fulfills outer core (5) then transitory layer (6), while the consecutive basaltic belts (fig. 2. unlined here) of the growing oceanic bottom (14) between continents (13) are chronologically equivalent to sequential geospheres of the mantle (7, 8, 9, 10, 11) up to the Earth's crust (12). *Inward growth* – a very fraction of the newly generated atom-molecular and energetic matter is going from the ultrathin transformable geo-spherical kernel (4), crossing a narrow melted peripheral zone of the outer core and the inner transition layer (3), and is added to the inner core (2), up to the inner-inner core (1). The long upward arrows starting from the core-mantle boundary (Layer D") point to the motion as mantle plumes of the liquid, potential magmatic matter newly generated from core kernel. So, the entire Earth is affected by the growing process. (Author).

Ultradense globules from a big blast
triggering the planet-formation

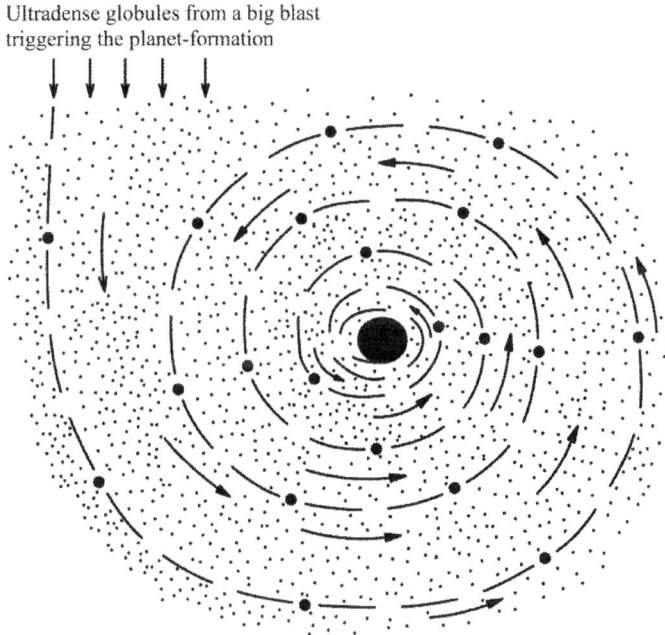

Figure 7. Scheme about ultradense globules inside of the gaseous. dusty cloud of the initial solar system
The ultra dense globules, **originated from supernova-like explosion,** inciting the planet formation process, are accreted to each other to become gravitational centers and potential dynamic cores of solar system bodies. (Author)

TEMPERATURES OF SUN'S ATMOSPHERE STRUCTURE

Figure 8 A views on temperatures of the Sun's lower *atmosphere layers*, **between photosphere and corona.**
Here in the figure is schematically shown that the photosphere's (Sun's surface) temperatures is about 6,000 K, when in the lowest corona in the height of 5500 Km it is above one million. Such phenomenon points directly to earth core structure, where highest temperatures ought to be outward of the core kernel, and toward the solid mantle boundary are falling very slowly , while inward, toward the inner core are falling very fast, and the atom-molecular mass is soon stiffened to inner core boundary.

Further, seismic studies unequivocally determine that earthquake energy comes from Earth's outer core and that the core activity and earthquake events are influenced by the Sun's cycles. There is an apparent correlation between earth seismicity and sunspot cycles. It is observed that earth magnetic field is disturbed by solar wind and also by earth seismic activity around the epicenter, some hour before earthquake shock. (Author)

80

www.ingramcontent.com/pod-product-compliance
Lightning Source LLC
Chambersburg PA
CBHW070844180526
45168CB00002B/956